'*Trans America* places the recent conversation about trans issues in its historical context, in impressive depth. Sweeping across the twentieth and twenty-first centuries, Barry Reay provides an accessible yet comprehensive guide to the important people, places, and trends, in the US and beyond – ideal for anyone who wants to understand what came before the "Transgender Tipping Point".'

Juliet Jacques, author of *Trans: A Memoir*

'The richly varied nature of the current trans movement is so beautifully explored and uncovered in Barry Reay's new book. A pleasure to read.'

Fayette Hauser of The Cockettes

'This is an admirable contribution to trans history by a highly respected scholar. It is a story of shifting categorizations, often highly medicalized and limiting, but above all a narrative of agency as trans people pushed definitions to the limit, bent them, and broke them, and increasingly spoke for themselves in a powerful, if not always singular, voice. It's a major achievement and deserves to become a classic.'

Jeffrey Weeks, London South Bank University

'This book is of very high quality. Reay is a major scholar in the field and writes with great authority and assurance.'

Thomas Laqueur, University of California at Berkeley

Trans America

Trans America

A Counter-History

Barry Reay

polity

First published in 2020 by Polity Press

Polity Press
65 Bridge Street
Cambridge CB2 1UR, UK

Polity Press
101 Station Landing
Suite 300
Medford, MA 02155, USA

ISBN-13: 978-1-5095-1178-5
ISBN-13: 978-1-5095-1179-2 (pb)

A catalogue record for this book is available from the British Library.
Library of Congress Cataloging-in-Publication Data

Names: Reay, Barry, author.
Title: Trans America : a counter-history / Barry Reay.
Description: Cambridge ; Medford, MA : Polity, 2020. | Includes bibliographical references and index. | Summary: "A history of trans before the "trans moment"-- Provided by publisher.
Identifiers: LCCN 2019043752 (print) | LCCN 2019043753 (ebook) | ISBN 9781509511785 | ISBN 9781509511792 (pb) | ISBN 9781509511822 (epub)
Subjects: LCSH: Transgender people--United States--History. | Gender identity--United States--History.
Classification: LCC HQ77.95.U6 R43 2020 (print) | LCC HQ77.95.U6 (ebook) | DDC 305.3--dc23
LC record available at https://lccn.loc.gov/2019043752
LC ebook record available at https://lccn.loc.gov/2019043753

Typeset in 10.75 on 14 Adobe Janson by
Servis Filmsetting Ltd, Stockport, Cheshire
Printed and bound in Great Britain by TJ International Limited

For further information on Polity, visit our website:
politybooks.com

Contents

Illustrations

Introduction

Trans seems to be everywhere in American culture. Yet there is little understanding of how this came about. Are people aware that there were earlier times of gender flexibility and contestability in American history? How well known is it, say, that a previous period of trans visibility in the 1960s and early 1970s faced a vehement backlash right at the time that trans, in the form of what was then termed 'transvestism' and 'transsexuality', seemed to be so ascendant? Was there transness before transsexuality was named in the 1950s and transgender emerged in the 1990s?

This book explores this history: from a time before trans in the nineteenth century to the transsexual moment of the 1960s and 1970s, the transgender turn of the 1990s, and the so-called tipping point of current culture. It is a rich and varied history, where same-sex desires and identities, cross dressing, and transsexual and transgender identities jostled for recognition. It is a history that is not at all flattering to US psychiatric and surgical practices.

There are competing narratives in trans history. Some have maintained that convictions of gender dislocation have always existed; this was claimed in *True Selves* (1996), the popular guide to transsexuality recommended by Jennifer Finney Boylan when she declared her transition to her academic colleagues: 'one indisputable fact remains: transsexualism exists and has always existed'.[1] The authors of *True Selves* were in good company. 'The historical records make it very clear that transsexualism has been a human problem since the most ancient times', wrote the wealthy, female-to-male transsexual Reed Erickson in his foreword to the classic *Transsexualism and Sex Reassignment* (1969).[2] For Max Wolf Valerio, a former radical feminist, 'People like me have always existed, in every era, on every continent.'[3] Yet this is not the case. As this book will explore, transgender does not float free of historical or cultural context.[4]

For others, far from 'always' existing, transsexuality was a late-twentieth-century phenomenon. As Catherine Millot once put it, there is a sense in which there was no transsexuality before experts like Harry Benjamin and Robert Stoller 'invented it'.[5] Although Joanne Meyerowitz's influential book on the subject has charted individual and sporadic instances of surgery and experimental sex modifications in Europe and (more rarely) in the USA from the early twentieth century, she effectively began her story with the intense publicity surrounding the sex-reassignment surgery of Christine Jorgensen in the 1950s: 'Ex-GI Becomes Blonde Beauty'.[6] Transsexuality, a category that had once not existed, quickly became a widely recognized term after it had been named and described in Benjamin's *The Transsexual Phenomenon* (1966), Richard Green and John Money's edited collection *Transsexualism and Sex Reassignment* (1969), and Stoller's *The Transsexual Experiment* (1975).[7] Before that, those who experienced gender disjunction would invariably have explained those feelings in terms of homosexual or heterosexual transvestism – such was the rapid movement of sexual classification.[8] Over the next ten years, the US national picture changed from one of no significant institutional support for transsexual endocrinology, therapy, and surgery to a situation where, by 1975, major medical centres were offering treatment and many transsexuals had been provided with surgery.[9]

One of the notable aspects of trans history is the rapid shift in sexual and gender configurations.[10] The transgender community emerged in the 1980s and 1990s, more sexually and gender diverse than the older transsexual community (which it incorporated) and less wedded to medical intervention.[11] When Anne Bolin published her study on male-to-female transsexuals in 1988, stressing surgery ('There are no halfway measures. If one is a transsexual, then pursuit of surgery accompanies one's transition'), it was in that period of movement from transsex to transgender – and already seemed dated.[12] By 2008, on the other hand, Walter Bockting was explaining that there was 'no one way of being transgender': 'Feminizing and masculinizing hormones and genital-reconstructive surgery are no longer two steps of one linear process of sex reassignment . . . Clients no longer necessarily need surgery to live and be recognized in the desired gender role.'[13] Trans surgery too – for wealthy trans women at least – has shifted from an emphasis on 'the genitals as the site of a body's maleness or femaleness' to an increased focus on the face as a site of true sex: moving from genital reconstruction surgery to facial feminization surgery.[14]

The category transgender includes people who want to create and/ or retain characteristics of both genders and who see themselves as neither or both male and female; significantly, other pieces by Bolin in the 1990s argued for far more gender flexibility.[15] The most recent large-scale survey of transgender people has discovered a vast range of different self-identity descriptions among those in the survey who classified themselves as 'other' or 'transgender', the more common self-descriptions including genderqueer, androgyne, and bi-gender.[16] *Trans/Portraits* (2015), which contains short testimonies of the experiences of a spectrum of American trans individuals, includes an array of trans masculinities and femininities, as well as those who identify as nonbinary, agender, and gender queer.[17] Dakota, who was agender, said that they were 'a sort of subset of genderqueer, in that I feel like I don't really have a gender at all. I don't feel male or female. I have elements of both sexes, or maybe neither.'[18] In short, there is a new awareness of the 'diversity of transgender experience'.[19]

We are now past the moment when the inaugural 2014 issue of the new academic journal *TSQ: Transgender Studies Quarterly*, itself indicative of the shift, could refer to the 'postposttranssexual'.[20] It is the era of trans*.[21] Transgender is considered too limiting, still connoting a gender binary. The asterisk in trans* indicates more openness, 'greater inclusivity of new gender identities [though even the notion of *identity* may be too restrictive as we will see later in this book] and expressions . . . such as gender queer, neutrios, intersex, agender, two-spirit, cross-dresser, and genderfluid'.[22] Aren Z. Aizura opts for 'gender non-conforming'.[23] More crucially, these terms do not necessarily reflect those used by trans people to describe themselves. They have often seen no ambiguity: that is an outsider perspective. Or they have embraced their blurring of conventional gender boundaries – for example, those who use the pronoun 'they' instead of 'she' or 'he'. Nonbinary has become a new category.[24] CN Lester prefers to be referred to as 'they'; and considers themself as 'outside of the gender binary', neither a man nor a woman.[25] *Aperture* magazine's 2017 visual homage to 'Future Gender' stresses gender as 'a playground'.[26] The androgynous, gender-fluid bodies of Ethan James Green's photographic portfolio *Young New York* (2019) capture the current moment perfectly.[27]

'Today trans is everywhere', wrote Jacqueline Rose in 2016.[28] There are trans-themed television series: Netflix's *Orange Is the New Black* (2013–19), Amazon Studios' *Transparent* (2014–17), and *Pose* (2018–), the last with significant trans participation in acting, directing, and the whole creative process.[29] There is an interest in transgender children that ranges from the 'superficially positive' to the downright hostile.[30] There is a developing trans fiction, aimed at young adults, clearly intended to educate non-trans readers and to support a trans audience.[31] There are trans celebrities: the very white Caitlyn Jenner of *I Am Cait* (2015–16) and *Vanity Fair* (2015) fame, and the black trans woman Janet Mock, with her best-selling memoirs and progressive advice about trans sex work and men who are attracted to trans women.[32] Trans women counsel non-trans women on their makeovers, reality television style.[33] YouTube has cleverly crafted – if highly idealized – visual records of trans self-fashioning, charting the respective effects of testosterone and oestrogen on trans man masculinity and trans woman femininity.[34] And the website has its own trans celebrities: Giselle Gigi Lazzarato, for example, with her 2.7 million YouTube subscribers.[35]

There is a comprehensive, trans, self-help guide, *Trans Bodies, Trans Selves* (New York, 2014), the trans equivalent to the iconic feminist text *Our Bodies Ourselves*.[36] There is a lavishly illustrated, colour-pictured guide to gender affirmation surgery, which does not spare the reader the lows as well as the highs of vaginoplasty and phalloplasty, and may not prove to be the best publicity for such procedures.[37] There are medical guides to assist health-care professionals in their treatment of trans patients, which, in contrast to earlier doctor–patient interactions (as we will learn), stress 'a therapeutic physician–patient alliance'.[38] Such humane principles have been comprehensively enshrined in the 'Standards of Care for the Health of Transsexual, Transgender, and Gender-Nonconforming People, Version 7' (2011), with its proclamation that being trans 'is a matter of diversity, not pathology', and in the World Professional Association for Transgender Health declaration (2018) that 'opposes all medical requirements that act as barriers to those wishing to change legal sex or gender markers on documents'.[39]

There are foundational *Transgender Studies Readers*, representing both Transgender Studies 1.0 and Transgender Studies 2.0.[40] There is a new transgender studies textbook, written by a nonbinary trans academic, intended for use by high-school and college students, and with significant input from trans contributors, including a section in each chapter called 'writings from the community'.[41] There is an anthology of trans poetry and poetics: 'Strange that you'd let me / give birth to my own body / even though I know I've always been / a boy, moving / toward what? Manhood?'[42] There are trans archives. The Transgender Archives at the University of Victoria, in Canada, is a relatively new archive (2011), formed out of the collection of Rikki Swan and the papers of Reed Erickson.[43] Cyberspace provides the scope for 'transgender history to be provoked, recorded, disseminated, accessed, and preserved in ways untethered from traditional, offline, and analog practices of history'; the curated Digital Transgender Archive is a most impressive demonstration of that very potential.[44] The Tretter Transgender Oral History Project of the University of Minnesota provides nearly 200 moving-image oral histories online.[45] There is a growing portfolio of trans photography: most recently, Zackary Drucker and Rhys Ernst's moving catalogue of a trans/trans relationship, and Mark Seliger's beautiful images of trans masculinity and femininity, and those in-between – 'endless possibilities of potential selves', in Janet Mock's words.[46] Vice.Com has set up the online Gender Spectrum Collection, providing free stock photographs of trans and nonbinary models (taken by Drucker) to increase the visual presence and enhance the media representation of those 'beyond the binary'.[47] See Illustration 1. Although it has just stopped publication, for ten years trans men had their own, genuinely innovative, magazine, *Original Plumbing*, edited by Amos

1   *A transmasculine doctor in front of his computer.*

Mac and Rocco Kayiatos, which, both visually and in prose, shows the sheer range and vibrancy of trans male culture.[48] See Illustration 2.

Hence, it has become possible to ask, 'Is Pop Culture having a Trans Moment?'[49] *Time* magazine cover stories can proclaim a 'Transgender Tipping Point' (with the black, trans woman Laverne Cox on its cover) and 'Beyond He or She'.[50] The *National Geographic*, no less, has had a special edition on 'The Shifting Landscape of Gender'.[51]

I will be using the literature of psychology, psychiatry, and modern surgery among my source material. But that does not mean that I have been captured by what is usually called the medical model, where trans is viewed through the lenses of the medical and psychiatric experts, the gatekeepers of transition. Some trans advocates, as we will see, are deeply suspicious of such influences; others have opted to work strategically within the system.[52] The trans community has long been divided on such issues.[53] On the one hand, the medical model provides (some) access to health care and (as a legitimizer) to legal advocacy, even if many of those involved do not really believe in the paradigm. On the other hand, it is resisted because it not only pathologizes but also privileges a particular kind of transgender, excluding more flexible forms of transness as well as those (the majority) precluded by poverty.[54] As Riki Lane expresses it, the 'tension between seeking approval for treatment and resisting pathologization is a defining characteristic of the relationship between clinicians and TGD [trans and gender-diverse] people, both as individuals and as a social movement'.[55]

Obviously, the medical model has framed discussion and shaped the lives first of transsexuals and then of transgender people; it has determined the rules, the parameters, the gates to treatment, and even self-perception. Austin Johnson's labels 'hegemonic' and 'normative' are entirely appropriate.[56] The sociologist Myra Hird was horrified by the attitudes of psychiatrists, physicians, and psychologists when she attended a gender identity conference in 2000, including 'highly stereotyped notions of gender' and the continued framing of transsex (and homosex) as pathology.[57] Many commentators have pointed to the persistent gender essentialism and heteronormativity of the paradigm still present in the regime of DSM-5.[58]

Yet, despite this dominating role, there has still been room for trans

OP

The Issues Issue

TRANS MALE CULTURE
THE ISSUES ISSUE • NUMBER 20 • 2019 • USA $10

2   Original Plumbing, *Issue 20, featuring Amos Mac and Rocco Kayiatos.*

agency, evidence of what Dean Spade has termed 'a self-conscious strategy of deployment of the transsexual narrative by people who do not believe in the gender fictions produced by such a narrative, and who seek to occupy ambiguous gender positions in resistance to norms of gender rigidity'.[59] Judith Butler once referred to San Francisco's 'dramaturges of transsexuality', who coached trans men in the gender essentialism which they did not personally hold – yet needed when they approached the psychiatrists and doctors who were the gatekeepers to the sought treatment.[60] 'I braced myself for a conversation where not adhering to stereotypes and clichés could undo this whole plan', the British trans woman Mia Violet recalled of her encounter with her therapist in the 2000s. 'I recited my history of gender dysphoria on cue.'[61] She carefully avoided complicating the expected narrative.

What clinicians took for patient duplicity could be interpreted as trans agency – as in the case of the famous Agnes, discussed in a later chapter. L. M. Lothstein, the psychologist at Case Western Reserve Medical School in Cleveland, whom we will also encounter later, held group therapy sessions in the 1970s in which patient power was evident. Some black trans women brought their street alliances (forged in sex work) into the clinic, where it became black patient versus white clinician. One, Ann, 'argued that the real experts on transsexualism were the patients and that the therapists were learning a lot about them via the group therapy'. She claimed that therapists could be 'bullied into recommending all patients for surgery'.[62] When a surgeon was invited in to show slides of gender reassignment, 'the group focused on the "ugliness" of the constructed vagina'.[63] In a later study, Lothstein and his team claimed that such therapy revealed material that had been 'denied' and 'falsified' in earlier evaluations, again evidence of patient initiative.[64]

Elroi J. Windsor has outlined the strategies (apart from submission) available to trans men when negotiating therapy: what Windsor terms 'manipulation' (choosing sympathetic therapists, and/or seizing back the initiative in the patient–therapist interaction), and 'resistance' (avoiding therapy, challenging diagnosis, walking away when the therapy does not suit). There are overlaps between categories, but the essential point is that, other than merely just 'doing what needed to be done' on the therapist's terms (which was also a tactic), trans men could operate within the medical model.[65] Readers should afford me the comparable ability to work the sources analytically, to read against the grain, rather than assume that I am the prisoner of a literature of which I am very critical anyway.

This will apply, too, with the discussion of surgery, which will recur in the pages that follow. We will see that many trans people eschew such accounts because they objectify and pathologize the trans body and pander (again) to the medical model. In his account of his trans journey, Nick Krieger consciously edited out descriptions of the immediate results of his top surgery in an effort to avoid a 'trans narrative cliché'.[66] Yet, either in its practice or in its absent presence (its denial), surgery has always been part of trans history.[67] As Eric Plemons frames it, 'I am an ethnographer of trans- surgical practice not because surgery defines us as trans- people but because it is so very important to so many of our lives.'[68]

We have to be wary of essentializing categories. Just as we should avoid subsuming transvestism under transsexuality, we should resist transgender as a master category for all aspects of trans history: the danger of the *Transgender Studies Readers* is that they may do just that. When Megan Davidson interviewed over 100 transgender activists in 2004 and 2005, well into the second decade of the transgender turn, she found conflict as well as shared values.[69] There were those for whom the medical model of transsexuality, with its binary and surgical certainties, was imbricated in their sense of self. Then there were those for whom fluidity was the key. The former sometimes saw the latter, especially those self-identifying as gender queer, as the province of white, privileged, college students. Davidson encountered an activist who clearly resented what they called the 'girl in a tie with a crew cut who now feels male and yet is not willing to manifest it other than [with] a tie and a crew cut'.[70] Raewyn Connell's deft history of transsexual women for a feminist readership demonstrates both an awareness of the emergence of transgender and her own preference for transsexuality as the more meaningful category, presumably because it best fits the centrality of the body to that history.[71]

Something strange is happening in some strands of trans studies: the erasure of much of trans history. Of course, historical frames of

reference vary. For Zackary Drucker, one of the current trans generation, the mid-1990s were formative, and she spoke of discovering the
words 'queer' and 'transgender' as a 'fourteen-year-old queer youth'.
Kate Bornstein was her 'gender pioneer'. But Bornstein, Zackary's
inspiration, had different influences and perspectives, other historical
reference points: Christine Jorgensen, Lou Sullivan, Tula's 1982 book
I Am Woman.[72] Writing in the early 1990s, Gordene Olga MacKenzie
identified the influence of the TV talk shows – mainly negative – on
trans 'coming out'.[73] For Rhyannon Styles, on the other hand, history
is compressed even more. Her inspiration, as a gay club kid, was reality
television. Before that, 'Men could only be women in pantomimes, or
when using drag to entertain'![74]

The most recent trans generation, of course, turns to the Internet,
to varied online communities, Gaming, Google, Facebook, Twitter,
Tumblr, and YouTube.[75] Tiq Milan has said that in the early 2000s
he thought that he was the only 'Black trans man in existence' until he
found a Yahoo discussion group.[76] 'Computer games were my mirror',
writes Shane McGriever, a trans boy, 'showing me the truth of myself
while giving me the purest escape from truth'.[77] For Harlow Figa, it
was YouTube's trans male vloggers ('up to ten hours a day') who were
his big influence: 'I learned how to speak about my transness through
YouTube.'[78] The queer, gender-nonconforming, and trans youth at the
drop-in centre studied by Mary Robertson found their sexual scripts
on Google, and in *anime* and fan fiction.[79] Not surprisingly, Genny
Beemyn and Susan Rankin's survey of nearly 3,500 transgender people
has argued that the Internet was crucial to transgender identity work
among the younger transgender participants.[80]

But, whatever the favoured medium, narrative, or cited forerunner,
the tendency has been to obscure what this book will argue was a
contested and troubled – even provisional – past. In Drucker's representation, the 1960s seem lost in the mists of time: 'For the 1960s, that
was so forward thinking.'[81] For genderqueer, nonbinary Jacob Tobia,
the 2000s – inconceivably, given all that you will read in this book –
provided no language to describe their genderless feelings, and 2009 is
almost ancient history: 'no one knew who Laverne Cox was yet (can you
imagine?)'.[82] Or take the historical introduction to *Vanity Fair*'s 2015
special edition, *Trans America*, that denies any 'smooth continuum'

from trans rejection to acceptance, yet which demonstrates the precise opposite by moving quickly to what it terms the 'sustained high' for transgender in contemporary US culture and to the celebrity trans promoted by that magazine.[83] Lest it be argued that these are examples of popular rather than academic culture, consider Jack Halberstam's recent book *Trans** (2018), which, apart from a discussion of 1970s feminism, has almost nothing from the period before the 2000s.[84] Of course CN Lester must be excluded from my criticism, for they have read widely in the historical literature and are thoughtful about the value of the past for the trans community: 'What I have learnt about our histories shows me that the gendered bars and limits placed around us need not be permanent.'[85] Similarly, many of the contributors to the edited collection *Trap Door* (2017) are committed to recovering a useable trans history.[86] But they are the exceptions that prove the rule.

When did this neglected history actually begin? Was it in the 1950s as already intimated? Or does this Jorgensen-inspired focus on those years distort a longer story? Julian Gill-Peterson has convincingly argued for 'displacing the 1950s as a default starting point for trans history'.[87] If it is possible to think of heterosexuality before heterosexuality, and homosexuality before homosexuality, why not think of transgender before transgender?[88] What is the history of trans feelings, tendencies – it is difficult to find the right term – before transsexuality and transgender were named in the second half of the last century? How useful is it to claim transsexual subjectivities for the late nineteenth and early twentieth centuries? Chapter 1, 'Before Trans', deals with these issues.

Chapters 2 to 4 examine the so-termed transsexual moment. Janice Irvine, one of the most perceptive observers of the twentieth-century historical sociology of sex, has written of transsexuality's 'widespread public and professional acceptance' by the 1970s, 'an accepted syndrome, buttressed by a vast medical armamentarium of research, publications, and treatment programs'.[89] But how seamless, really, was the triumph of transsexuality in the 1960s and 1970s? Chapter 2, 'The Transsexual Moment', discusses this ostensibly successful establishment of a new medical diagnosis and entity, arguing for the importance of cross-dressing (then known as transvestism) during this period of trans history. There is a case that the rather more fixed definitional qualities of the earlier 1960s and 1970s regime of transsexuality were necessary to establish a new category and to distinguish it from homosexuality and transvestism. However, we will see in Chapter 3, 'Blurring the Boundaries', that this sexual certainty masked a world of far more ambiguous alliances and practices. Chapter 4, 'Backlash', deliberates a neglected aspect of trans history, a period of intense critique right at the point where transsexuality had seemed to have become established.

Chapter 5, 'The Transgender Turn', considers the shift from transsexuality to transgender, and it assesses claims about the speed with which transgender has become established in the American cultural psyche. How, and in what ways, has that shift occurred? Has there been both a 1990s turn and a 2010s tipping point? Is trans culture really experiencing a cultural high?

Categories like transvestite, transsexual, transgender, and trans itself are good to rethink US history, but this book will demonstrate that it is the slippages and overlaps between these types that can be the most informative. As most dictionaries will explain, trans means across, beyond, over, and between; it can also denote change, transformation.[90] The history that follows will include those with transgender bodies before transgender emerged as a descriptor; those who cannot be categorized as either transvestite or transsexual; cross-dressers who modify their bodies but who are not transsexual; those who wanted to be homosexual rather than heterosexual after their bodily reconstruction; and those who consider themselves beyond classification. This book will locate and contest some of the more significant structural and conceptual weaknesses in trans history: the neglect of an important period of critique in transsexuality's early years; a claimed recognition of systems of technology and therapy and notions of sexual identity that I will suggest were far more tentative, contested, and fragmentary; and a neglect of other forms of trans expression both before and after the transsexual moment of the 1960s and 1970s. This book will attempt a new history of transsexuality and transgender in modern America.

1

Before Trans

Introduction

'When we first examined him [*sic*] we could not make up our minds whether we should send him [*sic*] to the men's wards or in with the women.' Such was a brief entry in the memoirs of the man who had been the Chief Surgeon at San Quentin, California's State Prison. 'After many careful examinations by many physicians', he continued, 'it was the consensus of opinion that Artie had been born a normal male child, and that some skilful surgeon, for reasons unknown, had operated . . . and turned him [*sic*], to all outward appearances, into a woman.' The memoirs are from 1940, so the recollection is from any time between then and (going backwards) 1913, when Leo L. Stanley took up his post at San Quentin. They are from the period before the transsexual moment of the 1960s and 1970s, when what is now called gender reconciliation became a feasible option for people like 'Artie'. The surgeon's description certainly hinted that some form of surgery had occurred, even if it was removal of the penis and testicles rather than any attempted vaginal reconstruction. 'Artie' was 'completely asexualized'. There had been attempts too at facial reconfiguration: 'Scars dotting his [*sic*] face were evidently the result of attempts to destroy the beard by electrolytic needle. What medical brute did such a thing, and why, we shall probably never know.'[1]

Leo L. Stanley's cryptic chronicle of attempted gender modification is typical of so much of the historical evidence in the period before transsexuality and transgender were named. One might assume that Stanley would have been attuned to the varieties of genital surgery, given his (notorious) eugenicist medical experiments with sterilization and testicular implants.[2] With his oversight of thousands of inserts of testicular substance, and experimentation with transplanting the testicles of executed prisoners, he should have known an absent testis when he did not see one.[3] However, he claimed that he was perplexed at first sight of 'Artie'. 'I thought he [*sic*] was a true case of dual sexuality [intersex as it would come to be called] . . . Leading physicians from the nearby cities came to examine Artie. It took a corps of them to determine if he [*sic*] were male or female.'[4]

As far as we are aware, the inmate 'Artie', our person of interest, left no historical traces other than these medical/mediated ones. It is possible that there are more detailed case notes in Stanley's archive at the California Historical Society in San Francisco, not obvious from a quick survey of its guide.[5] But, as things stand, the person known only as 'Artie' does not speak to us directly. Apart from the fact that they were eventually perceived by Stanley as a neutered male, we have no idea what their female name was. In keeping with modern trans sensibilities, we should probably attribute womanhood, yet we have no way of ascertaining how 'Artie' saw themself at that particular moment in San Quentin – whenever that was. Did they identify as female or male or neither? Stanley claimed that 'Artie himself [*sic*] was uncertain' and begged the Chief Surgeon to 'reinstate him [*sic*] by operative means to either male or female status'.[6] This did not occur; 'there is no hope of making a normal being of him [*sic*]. Trained only in bisexual perversion, syphilitic, and undoubtedly insane, what chance has this victim of human bestiality?'[7] Stanley's summary was harsh: 'A moronic monster, he [*sic*] could only jabber filth. He [*sic*] leered in answer to our questions and made obscene replies. Evidently, he [*sic*] knew nothing of his [*sic*] origin or sex, and only some of the most shocking adventures of his [*sic*] life were remembered.'[8]

What do we make of such cases? Can we get beyond the medical or disciplinary case study, the moral judgements, the objectification and victimhood? Was gender modification possible before the well-publicized cases of the 1950s? Are there other Arties, and, if so, are they part of trans history? Can we even think of trans before trans? While not assuming that transsexuality, transgender, and trans have always existed, what is their prehistory?[9] Some promising investigations have been 'trans-ing' (Clare Sears's term) the history of nineteenth- and early-twentieth-century America, so the issue is worth pursuing.[10] Emma Heaney has discussed what she calls the 'trans feminine' in some nineteenth- and early-twentieth-century texts, including some of the sexological works that are discussed in this chapter.[11] Yet her subjects too quickly become 'trans women' in the discussion.[12] Similarly, is Jay Prosser right to claim transsexual subjectivities for this prehistory?[13] Or is transhistoricity or 'trans*historicities' a better way of conveying – in the words of Kadji Amin – not some 'stable foundation' of transsexuality or transgender but rather 'a network whose nodes eventually shifted, were rejected, and fused with new elements to compose what we now know as "transgender"'?[14] How do we write the history of transgender before transgender?

The task is by no means simple. Rachel Hope Cleves's short biography of Frances 'Frank' Ann Wood Shimer, a nineteenth-century American teacher, once identified as lesbian, has posed five contemporary ways of making sense of his or her masculinity – with six if the more modern 'lens of trans studies' is invoked with Frank Shimer as a trans man.[15] But the danger is that, whatever authorial intent, this mere invocation, this naming, 'Six Ways of Looking at a Trans Man?', will foreclose the subtleties of historical reinvestigation.[16] Searching the US newspapers from the 1870s to the 1930s, Emily Skidmore has located sixty-five instances of 'individuals who had been assigned female at birth but [who] chose to live as male', many of them married to or living with women. Yet whether they should be described as 'trans men' (as Skidmore does) is completely different.[17] Or there is the question of what we do with the historical material. Over much the same period as that covered by Skidmore, Peter Boag has established the ubiquity of cross-dressing in the frontier West, both male-to-female and female-to-male, and Clare Sears has discussed the numerous San Francisco cases (nearly one hundred in all) of those who fell foul of the cross-dressing laws. She contrasts the popularity of on-stage cross-dressing performances (Julian Eltinge and Vesta Tilley) to the practice's criminalization on the streets.[18] But what to do with this omnipresence? As Boag writes of Portland's Harry Allen / Nell Pickerell, who in 1912 lived with a Seattle female sex worker and had reputedly caused the suicides of two women with whom he was involved (when they discovered his female origins):

> Did Allen don male clothes for economic reasons, or because *he* saw himself as a male, or because *she* wished to contest custom and law? Was Allen male or female, or a sexual invert? Did Allen's close relationships with women include a sexual component? Or did Allen simply have great sympathy for them, doing what he could with the limited means and opportunities available to him to assist them?[19]

Can we separate hints of sexual and gender identity from economic motivation? How many of the cases qualify as transgender before transgender?

Given that this is a book on American trans, it would have been logical to have discussed the nation's indigenous two-spirit people of the nineteenth and early twentieth centuries, formerly called 'berdaches' by Western observers, but known by a variety of local names by native peoples themselves: *lhamana* (Zuni), *winkte* (Lakota), *badé* or *bodé* (Crow), *ayekkwew* (Cree), *nadle* (Navajo), *alyha* and *hwame* (Mohave).[20] Remarkably widespread (indeed found in nearly 160 tribes[21]), they were the men and women who combined the gender roles (and often the clothing) of the opposite sex, and who were considered to be a third or fourth gender, 'not man, not woman'.[22] As one chronicler of the American West wrote in 1920 of a famous *lhamana*, the Zuni We'wha, 'She was a remarkable woman, a fine blanket and sash maker, an excellent cook, an adept in all the work of her sex, and yet strange to say, she was a man.'[23] However, in-depth historical analysis of the two-spirit people is difficult, given the layered interpretations of both colonial record-keepers, who classified berdaches as sodomites, and modern-day, queer Native American observers whose concept of two-spirit includes gay and lesbian as well as transgender identities.[24]

Sexology: Krafft-Ebing

As with so much modern sexual history, and along with the transsexual experts of the 1960s, we could start with the sexologists in Europe in the nineteenth century. Contrary sexual feeling or sexual inversion, terms that were synonymous with homosexuality in the nineteenth century, were conceived as a disjunction between an outer body and an inner soul. Karl Heinrich Ulrichs wrote in 1864 (talking about men) of a 'certain feeling of discomfort in one's own body, a certain dissatisfaction of the feminine soul with a body with the male form in which it is enclosed'.[25] He used the analogy of a hand in the wrong glove.[26] It sounds remarkably like the 'wrong body' narrative, which we will see is so influential in later transsexual histories.

The cases in Richard von Krafft-Ebing's *Psychopathia Sexualis* (1886) are especially interesting. Many were the case notes of medical or psychiatric consultation, forged in the interaction and negotiation of expert and patient and then classified according to the taxonomy of the medical scientist. However, others were unsolicited. Homosexuals read and responded to Krafft-Ebing's influential chronicles, for example, and, in turn, became incorporated into the next edition.[27] Although we should not minimize the power of sexology in shaping discourses and mapping out the parameters of sexual subjectivities, it was never a simple case of imposition. Harry Oosterhuis has argued that Krafft-Ebing's theoretical perspective was affected by the input of his patients; the influences were certainly not one-sided.[28] As Ivan Crozier has explained, the sexual categories the sexologists produced were 'manifestations of power' and a mapping of normality, yet there was interaction between authority and subject: 'People have an awareness of their own subjectivities, which is how they react to discourses about them (by accepting them, by understanding themselves in these scientific terms, by resisting them, by actively ignoring them).'[29] The sexological studies are important not merely for their categorization and analysis of forbidden desires but for the case histories that they contain.

Krafft-Ebing believed that sexual desire was gendered and comprised of both bodily and psychical elements. 'If the sexual development is normal and undisturbed', he wrote, 'a definite character, corresponding with the sex is developed. Certain well-defined inclinations and reactions in intercourse with persons of the opposite sex arise; and it is psychologically worthy of note with what relative rapidity each individual psychical type corresponding with the sex is evolved.'[30] In other words, in 'normal' sexual development males assumed male bodies and matching psyches and desired females accordingly. Females acquired female mental and bodily characteristics and desired males. Homosexuality represented a disjunction of these alignments. He insisted on the gendered interaction between the physical and the psychological. This was the process that he referred to as eviration (the feminization of men) and defemination (the masculinization of women), involving 'deep and lasting transformations of the *psychical* personality'. With men, the patient

> undergoes a deep change of character, particularly in his feelings and inclinations, which thus become those of a female. After this, he also feels himself to be a woman during the sex act, has desire only for passive sexual indulgence, and, under certain circumstances, sinks to the level of a prostitute . . . The possibility of a restoration of the previous mental and sexual personality seems, in such a case, precluded.[31]

One can see in these descriptions of male–male sexual desire (the references to sex work excepted) the germs of descriptions of later male–female transsexuality.

It is significant that homosexuality, or rather antipathic sexuality, is defined as loss of masculinity or effeminization. Case 128 in *Psychopathia Sexualis* involved one Sch., aged 30, a physician, who constantly related his life in terms of a homosexuality revealed as effeminacy: 'I developed a desire to move in ladies' society . . . was interested in toilettes and such feminine things';[32] 'I am effeminate, sensitive, easily moved, easily injured and nervous.'[33] Although he claimed that his manner and appearance were 'masculine', he cited his love of the theatre and the arts and his frivolity and lack of interest in manly pursuits as evidence of his womanlike qualities. He paid soldiers for sex, and lived with a succession of older men: 'we lived like a pair of lovers. I was the wife and was formally courted by the lover.'[34] Krafft-Ebing's diagnosis was that the man's desire for the 'passive role' and 'passive pederasty' are so strong that 'this urge extends its influence to the character, which becomes feminine to the extent that Sch. prefers to associate with real *women*, is more and more interested in feminine pastimes, and even resorts to make-up and other titivation in order to revive his fading charms and make "conquests".[35]

Krafft-Ebing held that, at its most extreme, acquired homosexuality could exert sufficient force on the psyche to create total sex delusion. Hence Case 129, a Hungarian-born physician, who 'felt exactly like a woman' after he had taken particularly potent hashish.[36] But the drug was merely a trigger to a permanently experienced state:

> I feel like a woman in a man's form; and even though I often am sensible of the man's form, yet it is always in a feminine sense. Thus, for example, I feel the penis as clitoris; the urethra as vaginal orifice, which always feels a little wet, even when it is actually dry; the scrotum as *labia majora*; in short, I always feel the vulva.[37]

The balding and grey-bearded Hungarian claimed (in 1893) that for three years he had 'never lost for an instant the feeling of being a woman'; he experienced menstrual discomfort, without actually bleeding; dreamed as a woman; saw his marriage as two women living together, 'one of whom regards herself in the mask of a man'; felt like a woman clothed as a man when he went about his everyday duties as a physician; sometimes cross-dressed, while alone, to feel more comfortable; and worried that, because his anus felt 'feminine' it might lead him into pederasty.[38] He concluded, again with shades of what would later be termed gender dysphoria, that 'I have a desire to be sexless, or to make myself sexless. If I had been single, I should long ago have taken leave of testes, scrotum and penis.'[39] The autobiography had been sent to Krafft-Ebing in response to an earlier imprint of his work and had been incorporated in a later edition. The letter accompanying the text, medical man to medical man, hoped that the psychiatrist would be interested in 'how a masculine being thinks and feels under the weight of the imperative idea of being a woman' – and Krafft-Ebing was indeed interested.[40]

Emma Heaney fastens on Case 129 as an example of 'a trans feminist theory of embodiment': 'She *is* a woman who *has* female genitals.'[41] Jay Prosser reads this case as an early located form of transsexuality.[42] He cleverly suggests that, rather than seeing sexual inversion as equivalent to homosexuality, it should be equated with transgender – 'sexual inversion *was* transgender' – and that homosexuality was but a strand or 'aspect' of this wider category.[43] The 'dynamic of transgender', he writes, 'of gender identifications that cross ("trans") at angles to bodily sex, is arguably sexology's main subject'.[44] However, Prosser's strategic switch mistakes the atypical for the standard. Most cases of sexual inversion were nothing like Case 129. It is significant that Krafft-Ebing saw this 'sense of a change of sex' or 'Change of Sex Delusion' as 'unique'.[45] Moreover, the sexologist interpreted Case 129 as a strand of 'homosexual feeling', 'perverse feeling for the same sex'.[46] For him, homosexuality was the master grouping.

Unsurprisingly, Krafft-Ebing refers to homosexuals as Urnings, evoking Ulrichs's idea of the female soul or psyche in a male body.[47] When he came to summarize the diagnosis of antipathic sexual instinct, he repeated the essential characteristic of the disjunction between psyche and anatomy. 'Every case of genuine homosexuality', he wrote, 'must be reduced to an abnormal sexual instinct which is diametrically opposed to the physical sex of the affected individual'.[48] This does not mean that there are no trans elements in this early conceived homosexuality – the Hungarian physician described them as his feeling that he was a woman in both body and psyche and with a 'feminine' interest in men.[49] Yet if Prosser is keen to link the sexology of the nineteenth century genealogically with the early sexology of transsexuality, surely it is significant that the medical men of the later period focused on the homosexual elements of transsexuality (defined in terms of the individual's gender assignment at birth rather than their perceived identity). Transsexuality in the 1950s and 1960s, we will see, was recognized in terms of its links to male and female homosexualities rather than with any transgender comprehension of sexual inversion. Prosser may be right to critique what he terms 'absolutist constructionism', the 'market theory' of transsexuality, which considers the phenomenon impossible until it was named in the mid twentieth century and hormones and surgery made transition feasible.[50] However, that does not mean that the alternative is some kind of transsexual essentialism. History shows *not* that transsexual desires are 'astoundingly consistent', but rather their inconstancy.[51]

Krafft-Ebing's case studies are notable for their range rather than conformity to a strict stereotype; but his overall representation of homosexuality, the weighting of his published cases, favoured the effeminate. Another of his categories was 'effemination', 'cases of completely developed inverted sexuality'.[52] In these men, like the evirates of acquired homosexuality, sexual desire permeated 'mental being'; 'the men, without exception, feel themselves to be females'.[53] As children, the boys liked to engage in girlish activities (another trope of transsexuality), playing with dolls and the like.[54] 'In homosexual intercourse effeminated man feels himself in the act always as a woman.'[55] Hence Case 149, B., a waiter, could only have intercourse with a woman if he thought of her as the man to whom he was really attracted.[56]

The final extreme of antipathic sexuality, a direct transition from all the other categories as Krafft-Ebing tellingly described it, was what he termed 'hermaphroditism', by which he meant not literal hermaphroditism, or intersex as it is now known, for the genitals were the only part of the body not affected by identification with the opposite sex.[57] In these extreme cases of homosexuality, the psychic identification with the female was accompanied by bodily change in the sense of 'the frame, the features, the voice, etc.', so that, as Krafft-Ebing described it, 'the individual approaches the opposite sex anthropologically, and in more than a psychical and psycho-sexual way'.[58] Hence, the body of von H. was 'rich in fat' and delicate and soft, with a mincing gait, effeminate manner, and traces of powder and paint.[59] For Krafft-Ebing, the logical extension of homosexuality was, quite literally, sexual inversion: both mind and bodily presentation conform to identification with the opposite sex.

In the well-pronounced cases of antipathic sexual instinct (effeminatio [male effeminacy] and viraginity [female masculinity]) the physical and psychical characteristics of inverted sexuality are so plentiful that a mistake cannot occur. They are simply men in women's garb, and women in men's attire, especially if they have full freedom of action. Psychically they consider themselves to belong to the opposite sex. We have seen female homosexuals in the army, and male homosexuals among the waitresses in restaurants. They act, walk, gesticulate and

behave in every way exactly as if they are persons of the sex they simu-
late. I have known male homosexuals who excelled women in wiles,
loquacity, coquetry, etc., etc.[60]

So the association of homosexuality with effeminacy received powerful
reinforcement from the sexologists, despite the more mixed messages
of their collected histories. The history of trans – in the sense of 'over
or opposite' as Magnus Hirschfeld might have put it – is closely linked
to the birth of the homosexual.[61]

The reader will have noticed that the focus has been on the male
'invert'.[62] 'Science in its present stage', wrote Krafft-Ebing, 'has but few
data to fall back on, so far as the occurrence of homosexual instinct in
women is concerned as compared with man'.[63] But he still had something to say on the subject:

> Careful observation among the ladies of large cities soon convinces
> one that homosexuality is by no means a rarity. Uranism may nearly
> always be suspected in females wearing their hair short, or who dress
> in the fashion of men, or pursue the sports and pastimes of their male
> acquaintances; also in opera singers and actresses, who appear in male
> attire on the stage by preference.[64]

The tropes were comparable to those discussed in relation to males.
The 'masculine soul, heaving in the female bosom' revealed itself early
in boyish play.[65] The female extreme was declared in the bodies of
those women whose only sign of femininity was their 'genital organs':
'thought, sentiment, action, even external appearance are those of the
man'.[66] Thus Case 162, a 26-year-old maidservant, had preferred boys'
toys and games when young, had lascivious dreams 'only about females,
with herself in the *rôle* of man', had cross-dressed as a man, and had
worked as a male servant.[67] Case 163 dressed like a man and 'felt as a
man towards women'.[68] Case 164 had 'engaged herself to a young girl
under the pretext that she was a man' and 'bewailed the fact that she
was not born a man'.[69]

But it was Case 166 that represented the most comprehensive
demonstration of 'inversion of the sexual instinct', and that has been
declared by Prosser as 'transparently transsexual'.[70] 'Countess V.',

Sarolta Vay, assigned female at birth, spent much of his adult life as Count Sandor Vay until prosecuted in 1889 by his father-in-law when he discovered that his daughter's husband was 'a woman'. Until then, the father-in-law stated, 'nobody had doubted his masculine sex'.[71] Vay's wife was allegedly deceived, and it transpired that the Count, as a Count, had had numerous trysts with women, including a previous marriage and separation.[72] The claimed lack of doubt was not quite true. Witnesses testified that in Vay's native Budapest 'everybody knew her [sic] and was used to seeing her [sic] as a fool'.[73] But in the spa where the couple met, the maintained masculinity was unchallenged, though hotel servants would later voice their gender suspicions.[74] Prosser treats Vay as transsexual (including the masculine descriptor), and, in conventional transsexual mode, emphasizes gender dysphoria over sexuality. He cites Vay's repugnance towards menstruation and masturbation as evidence that he did not (again anticipating transsexuality) want any reminders of his female body: 'a much more fundamental rejection of bodily sex is at work than can be explained through (homo)sexuality'.[75] But actually, sexuality was much in evidence in the case notes, not just in the references to a 'priapus'.[76] Vay had 'many *liaisons* with ladies', visited brothels, lusted over a cellmate, performed cunnilingus, and from their early teens experienced 'a trace of sexual feeling' for the female sex 'which expressed itself in kisses, embraces, and caresses, with sexual pleasure'.[77] It is true that Vay said that 'she [sic] never allowed herself [sic] to be touched on the genitals by others', but for the reason that 'it would have revealed her [sic] great secret' rather than because it would have reminded him of any despised female anatomy.[78]

It is a giant step, then, to Prosser's claim that gender inversion rather than same-sex attraction was sexology's main subject. There are, as we have seen, elements in the sexological cases that *cumulatively* contain the constituents of transsexuality, but more fully realized individual examples are rare. Case 129, we saw, was unique. Geertje Mak, Vay's historian, considers his case 'exceptional'.[79] Though we have detected components of what would come to be called transsexuality, they were not equivalent – nor limited – to transsexuality or transgender: they were necessary, but not sufficient, historical ingredients.

Sexology: Hirschfeld

If Krafft-Ebing's trans traces were linked to homosexuality in the prehistory of transsexuality and transgender, Magnus Hirschfeld shifted the focus to heterosexuality with his book *Die Transvestiten* or *Transvestites* (1910) – at least, on the face of it. He claimed that his seventeen case studies of cross-dressers, sixteen men (defined by assignment at birth rather than any claimed identity) and one woman (also based on attribution at birth), were a separate category, distinguishable from homosexuality and fetishism and sado-masochism – this categorization and separation was important to him. He gave a name, 'transvestism', to what would much later be called trans sensibilities. He maintained the importance of his project. It was fitting, he wrote, 'to give the new form a new name, a special scientific stamp'.[80]

The Transvestites certainly provided the makings of transsexuality. The connecting thread in the 'circle of people described here', Hirschfeld wrote, is 'the strong drive to live in the clothing of that sex that does not belong to the relative build of the body', and that the choice was by no means arbitrary 'but rather is a form of expression of the inner personality as a valid symbol'.[81] His transvestism involved more than clothing. Many of his transvestites (recall his focus on men) wanted to live some or all of their lives as women: 'They know all too well that a profound contradiction exists between their bodies and their souls . . . most of them wish that they had been born female.'[82] As one of his cases expressed it, 'My sexual wish is not to be the woman of the female impersonator, but rather my ideal would be, as a woman, to lead a genuinely physiological love life with a man.'[83] While all his subjects were attracted to clothing of the opposite sex (or their true sex in transsexual interpretation), Hirschfeld was well aware that what he was dealing with was 'not simply a matter of cross-dressing, but rather more of a sexual drive to change', and he contemplated calling the phenomenon 'sexual metamorphosis'.[84] When he said of his cases that they knew 'all too well that a profound contradiction exists between their bodies and their souls', he was talking about what would come to be called transsexuality, rather than transvestism.[85] Here we have the genesis of the distinction between the transsexual and the transvestite.

There are intriguing life stories that demonstrate that living life as the opposite sex could involve more than economic opportunism. Case 13, an Austrian 'man' in their forties who ended up in San Francisco working for a German Milwaukee newspaper, had moved around the world, from country to country, often dressed as a woman and engaged in women's occupations: a domestic servant, an embroiderer. They would share beds with female employees and had sex with at least one of them. They had frequently been discovered but never arrested. In San Francisco, they had at one point dressed in male clothing while working as a bookseller, but donned female clothing at home, where they were 'room mother' for a group of (female) dancers.[86] Describing themself as 'physically a man, mentally a woman', they said they wanted to marry a 'man-woman', a masculine woman, and live 'as a wife', abandoning 'hateful men's clothing'.[87]

Case 15, the only female-to-male transvestite, was a housepainter in East Berlin whose womanhood was discovered after they were accused of adultery with a married (though separated) woman. Hirschfeld saw them in 1907 when they were 27, married (to a man) and had children, but still had a 'burning wish to be a man, to go as a man, and to live as a man'.[88] Like the previous case, they had travelled the world, working sometimes as a woman and more often as a man – a coalminer, locksmith's assistant, butler, on a whaling ship, as a helmsman on a steamer. Although they said that their 'drive changes between both sexes', and were married to a man, much of their history consisted of relationships with women, because, as they put it, 'in the Bible it says, it is not good to be alone'.[89] There is reference to a jilted female fiancée, and they were discovered when, in their words, they 'began the craziness of looking for a[nother] fiancée'.[90] Life stories like this provide an indication of the sexual and gender intricacies concealed in the simple charge of cross-dressing.

Darryl B. Hill has argued that Hirschfeld's actual case studies did not present the neat separations that he emphasized in his classification.[91] They contain, for example, the elements of fetishism that would be denied by later transsexuality. Case 7, a 40-year-old former policeman, recalled early ejaculations while wearing women's clothing.[92] Case 9, a 37-year-old former officer in the American army, masturbated while cross-dressing: 'The yearning to feel totally like a woman also leads me to have coitus "with myself" [between the legs] using wax candles, cigars, and things like that.'[93] When the New York psychiatrist Emil Gutheil reworked Hirschfeld's data in the 1950s, he detected elements of homosexuality, sadomasochism, narcissism, scoptophilia (pleasure in looking), exhibitionism, and fetishism, some dominating more in particular cases, but all detectable in 'every case', and very different from Hirschfeld's unitary interpretation.[94]

Then there is the issue of his 'heterosexual' transvestites. Though Hirschfeld was well aware of the cross-gender attributions of homosexuality and expected to find a homosexual component in transvestism, he claimed that he had in fact found little link. '[W]hen the drive to activity came forward', he noted of his informants' sexual desires, 'in almost all cases it immediately directed itself, agreeing with the physical constitution, toward persons of the opposite sex. Almost all these persons put the thought of homosexuality out of their minds, many stating an instinctive loathing.'[95] Hence Prosser enrolled Hirschfeld in his distancing of the history of transsexuality from the history of homosexuality: 'his subject is not homosexuality; most of his subjects were not homosexual but heterosexual'.[96]

However, all those male-to-female transvestites who had sex with or desired women, and whom the sexologist saw as heterosexual, were rather strange heterosexuals. Hirschfeld said of Case 3, an artist in their forties, that their 'sexual drive was always directed towards the female; and intercourse is possible only with women. The thought of homosexual intercourse is repugnant to him [*sic*].' Yet Case 3 wished 'he [*sic*] had been born a woman . . . My yearning is not limited to women's costumes, but also extends itself to an absolute life as a woman.'[97] Case 4's desire was also exclusively directed at women, but the merchant, who was in their thirties, could not have sex with their wife unless they imagined themself as a woman.[98] Case 6 (as with many of Hirschfeld's informants) preferred intercourse with the woman on top, seen then as unmanly submissiveness: 'In order to have sex with my wife, she, at least, had to be wearing the clothes I would have liked to have worn.'[99] Case 13 preferred manly women as their lovers because they 'always make me feel like a woman'.[100] These lesbian-like 'heterosexuals' anticipated what transsexuality would come to call female homosexuals, because they were (really) women. Case 17, a 24-year-old lawyer, came closest to articulating this when they said, 'In general I feel attracted as a woman to women, and, if I were a real woman, truly would love the same sex.'[101] Such cases were not really heterosexual at all.

Even if we use Hirschfeld's definition of homosexuality – in terms of assigned-male-at-birth, male-to-female transvestites desiring men – it is clear that, as Hill has also discussed, several of his informants did not confirm to the claimed heterosexuality.[102] Case 3 yearned 'for a man'.[103] Case 9 wanted a man to penetrate them by force.[104] Case 12, a lawyer in their mid-twenties, 'came upon the idea of getting a man to comple-ment my desire'.[105] Case 4's biographical notes referred to 'the world of unlimited possibilities' beyond the world of 'heterosexual intercourse': 'Just writing about this now makes my penis . . . grow erect.'[106] The merchant fantasized about oral sex with their male friend: 'if my friend exposed his large penis to me and wanted me and I was excited, I would be willing to have intercourse with him'.[107] (Hirschfeld dismissed this as 'erotic fantasy' and said that the 'man' was no longer interested in such ideas.[108])

Had we used modern transgender meanings, of course, Hirschfeld's 'homosexual' transvestites would have been heterosexual – because they were really women desiring men.

Sexology: Ellis

The sexologist Havelock Ellis also wrote about and published case studies of transvestism, or what he first (1913) termed 'sexo-aesthetic inversion' and then (1928) called Eonism (after the gender-shifting his-torical figure the Chevalier d'Éon).[109] Like Hirschfeld, he emphasized the 'heterosexuality' of transvestism, again within the context of an initial interest in homosexuality. 'Many years ago, when exploring the phenomenon of sexual inversion', he wrote, 'I was puzzled by occasional cases I met with of people who took pleasure in behaving and dressing like the opposite sex and yet were not sexually inverted; that is, their sexual feelings were not directed towards persons of their own sex.'[110] The phenomenon, he explained, 'is one of erotic empathy, of a usually heterosexual inner imitation, which frequently tends to manifest itself in the assumption of the habits and garments of the desired sex'.[111] When he wrote 'Eonism', he rejected his earlier reference to 'sexo-aesthetic inversion' because he thought that mention of inversion was 'too apt to arouse suggestions of homosexuality'.[112] That his cases were heterosexual is not surprising; Ivan Crozier has shown that Ellis sought out only non-homosexual cases (in terms of the definitions discussed earlier).[113] His cases are mostly male-to-female too.

Ellis incorporated previous studies in what Crozier labels 'positioning rhetoric', referring to Krafft-Ebing's Hungarian doctor as 'the earliest full and scientifically described case'.[114] However, the interesting issue is: did this positioning result in any advance on previous work? Ellis is somewhat neglected by Prosser, but did his case histories (at the risk of being seen as teleological) bring us any closer to the transsexual moment?

In a sense, his case studies provide the same mixture as those of Krafft-Ebing and Hirschfeld: including those, as Harry Benjamin would later express it, who merely wanted to dress as women or wear some form of women's clothing, often with erotic motivations; those who were more committed to presenting or living as the other sex; and those who actually felt that they were members of that (alternative) sex – 'trapped' in a man's body was how Benjamin would describe such (wo)men.[115]

Thus, in his 1913 article in the *Alienist and Neurologist*, we have J. G., the 35-year-old, 'insatiate in his [*sic*] sexual desires', who had sex with their wife while both wore women's corsets.[116] A. T., an artist, aged 30, felt that they were a woman when dressed as one. They had tried to develop their breasts and said that they had 'almost reached the stage described as actual sexual inversion. When dressed as a woman, I am a woman, with all a woman's feelings and longings.'[117] In their account, relayed by Ellis, they told of liaisons 'with younger women and girls who were glad to find a male [*sic*] admirer who could indulge in unlimited lascivious caressing without always wanting to go always to the full length of actual connection'.[118] While they had sex with women while dressed as a woman, they had 'been assailed with', but resisted, the 'longing to give myself thus to a male instead of a female lover'.[119] Curiously, Ellis, *à la* Hirschfeld, denied any homoerotic component.[120]

'Eonism' provided more cases. T. S., a successful author in their fifties, had derived erotic pleasure from cross-dressing. Their published novels were assumed to be the work of a woman, and parts of one, they claimed, had been 'drafted while dressed and made up as a woman, often before the glass';[121] 'When I first began to dress as a woman, I was offended by the fact that it induced erection . . . but before very long . . . there was no disagreeable effect. I could entirely forget I am a man [*sic*].'[122]

Such cases were reminiscent of earlier sexology. But there were others that were closer to transsexuality. Ellis's 'most profound and complete' case of 'sexo-aesthetic inversion' was R. M., a scholar in their sixties, beset by 'the most passionate longing to be a woman'.[123] They felt their breasts changing and had earlier contemplated castration. They seemed, from their account, to be constantly examining their body and mind for signs of femininity, and described how they envied rather than desired beautiful women. R. M. actually used the term 'aesthetic inversions' to describe 'psychical affections' like theirs, which they saw as a feeling of being double-sexed: of having a strong affinity with women, rather than feeling that they were a woman. (They hinted that the work or experience of George Moore and Dante Gabriel Rossetti reflected similar tendencies.)[124] But they also described it as a feeling that one was born 'out of their sex'. And they denied homosexuality ('ordinary inversions'), given that the attraction was to women not men.[125] Ellis thought that R. M. demonstrated the insufficiency of transvestism as a category: 'The inversion here is in the affective and emotional sphere, and in this large sphere the minor impulse of cross-dressing is insignificant.'[126]

In 'Eonism', Ellis discussed the case of C. T., who was in their twenties and whose situation was complicated by tattooing and piercing fetishes. They indicated a growing desire to 'become more of a woman', including thoughts of castration: 'I know that I would be immensely happier if my sexual organs were removed. If I knew anyone who would perform the operation I should immediately have recourse to him.'[127] Finally, R. L., aged 48, actually spoke the language of later transsexuality, referring to 'the soul of a woman . . . born in a male body', of an early 'desire to be a girl', and of a feeling 'that I would, if I had the chance, be changed into a woman physically';[128] 'I would undergo a surgical operation if the result would be to give me a beautiful or attractive female form with full womanhood in a type that appealed to me.'[129] R. L. 'felt, and still feel, that my real self has had to be subjected to my physical self, my body'.[130] Ellis's article in the *Alienist and Neurologist* certainly prefigured the difference between what would come to be known as transvestism and what would become transsexuality:

> There are at least two types of such cases; one, the most common kind, in which the inversion is mainly confined to the sphere of clothing, and another, less common but more complete, in which cross-dressing is regarded with comparative indifference but the subject so identifies himself [sic] with those of his [sic] physical and psychic traits which recall the opposite sex that he [sic] feels really to belong to that sex, although he [sic] has no delusion regarding his [sic] anatomical conformation.[131]

Here – if under the umbrella of 'sexo-aesthetic inversion', and without any reference to the wish to modify any 'anatomical conformation' – is the essential distinction between transvestism and transsexuality. It is strange that Prosser has not enrolled Ellis more in his own narrative of the emergence of transsexuality.

A Transsexual Incubator

Yet, in a sense, all roads return to Hirschfeld. Not only did he introduce the term 'transvestite' (however it was comprehended) but he also recognized those committed to bodily modification. And he provided visual representations of the new identity – trans women in dresses and ball gowns, trans men in suits, naked bodies demonstrating the contours of the other (true) sex, manly women and womanly men – a strange mix of detached, medicalized photography (photography in the service of sexology), and more engaged imagery, presumably generated by the transvestites themselves, as David James Prickett, Kathrin Peters, and Katie Sutton have demonstrated.[132] In 1926, he used the term 'total transvestism' to describe 'those who want to transform not only their sartorial but also their biological appearance . . . These strive for a complete transformation of their genitalia . . . This means the elimination of menstruation by removing the ovaries for female transvestites, and for men castration. The number of cases is much greater than one had anticipated before.'[133] Thus Hirschfeld's Institute for Sexual Science became a focus for surgical and hormonal modifications in the 1910s and 1920s. See Illustration 3. Male-to-female transvestites worked as maids there. And Ludwig Levy-Lenz, who carried out surgery at the Institute in the late 1920s, recalled what would come to be called gender reconciliations:

> The first thing to be dealt with was depilation . . . elimination of hair on the face, which is a difficult process, for treatment with a needle takes a long time, and X-rays often cause severe burns. After that, the penis was amputated, then they demanded castration, and then an artificial vagina. I performed these operations at the Institute . . . and I did in fact succeed in forming an organ resembling the vagina – never have I operated upon more grateful patients.[134]

Female-to-male patients requested mastectomies. Levy-Lenz became an early plastic surgeon of facial modification as women wanted masculine features and men requested feminine features to bring their faces in line with their inner selves. He described such surgery as experiments – 'for I cannot call my first operations of this kind by any other name'.[135] There is a direct link between Hirschfeld, the Institute, and the work of Harry Benjamin in the 1950s. Benjamin knew Hirschfeld and visited the Institute frequently in the 1920s.[136] He also arranged for the German sex researcher to visit the USA in 1930 and hosted him while he was there.[137] Hirschfeld's Institute was, as Robert Beachy puts it, a 'veritable incubator for the science of transsexuality'.[138]

So Prosser is perfectly justified in claiming Hirschfeld for the prehistory of transsexuality: 'Hirschfeld's "transvestite" was pivotal to the discursive emergence of the transsexual.'[139] However, Hirschfeld's actual case studies – contra his own interpretations – chime with transgender rather than transsexuality. In the words of Katie Sutton, the introduction of Hirschfeld's term '*transvestit*' provided a word to label something closer to 'what would now be described as "transgender" experiences and identifications; from self-defined heterosexual male transvestites with wives and children . . . to biological females and males living permanently as the "opposite" sex'.[140] Significantly, when Hirschfeld's students published a posthumous textbook compendium of his writings in 1948, his original position had been modified somewhat. That transvestism was a category in itself was confirmed, but the strict link with heterosexuality had gone: 'From the comprehensive

3 *Transvestiten vor dem Eingang des Instituts für Sexualwissenschaft [Transvestites in front of the Institute for Sexual Science], 1921.*

data at our disposal we find that about 35 per cent of transvestites are heterosexual and an equal percentage homosexual, while about 15 per cent are bisexual. The remaining 15 per cent are mostly automono-sexual [autoerotic and fetishistic], but also include a small proportion of asexuals.'[141]

Sexology in America

This is a book about transgender in the USA, however. What were the sexological influences in America? Were there comparable case studies? One of the first was Lucy Ann Slater / Lucy Ann Lobdell / Revd Joseph Lobdell, the subject of an article in the *Alienist and Neurologist* in 1883, which cited Krafft-Ebing and another sexologist, Karl Friedrich Otto Westphal.[142] For the author of the case study, P. M. Wise, a physician at the Willard Asylum for the Insane in New York State, Lobdell was a disturbed 56-year-old widow who had long dressed and worked as a man and who, for a number of years, had lived with a woman 'in the relation of husband and wife'.[143] 'In nearly her own words', as Wise expressed it intriguingly, 'I may be a woman in one sense, but I have peculiar organs that make me more a man than a woman.'[144] The physician emphasized the sexual aspect of the case. Lobdell was an erotomaniac; it was a 'Case of Sexual Perversion'. 'I have been unable to discover any abnormality of the genitals', he continued, 'except an enlarged clitoris . . . She says she has the power to erect this organ in the same way a turtle protrudes its head – her own comparison.'[145] In an early sexological reference to lesbianism, Wise refers to 'Lesbian love'.[146] But more recent commentators think that transgender might now be a better description for someone who chose to live and love as a man.[147]

In the early days of American sexology, then, trans before trans fell under the rubric of perverted sexual instinct. Thus, a learned article in the *Journal of Nervous and Mental Disease* in 1883, familiar with the work of many European sexologists, and hoping to make it more accessible to their American colleagues, recopied cases of a 'woman' who 'had a great desire to be a man', a 'man' who dressed as a woman, a 'man' who 'feels himself more of a woman than a man', and another who 'used the pronoun "we" when speaking of women' and who 'felt an unconquerable instinctive desire to dress as a woman and to be one'.[148] All these cases of people who exhibited a 'yearning to belong to the opposite sex' suffered from a 'perverted sexual instinct'.[149] The Chicago physician J. G. Kiernan, who knew of both Lobdell and Ulrichs, wrote that he left the question of souls to theologians but was 'certain that a female functioning brain can occupy a male body, and vice versa'.[150]

'Psychical hermaphroditism' was how a Baltimore doctor, who had read Ulrichs, Krafft-Ebing, Ellis, and Westphal, categorized the case of a 39-year-old in 1897 who was strongly attracted to their 'own' sex.[151] It was not so much their sexual desires that attracted the doctor's interest, but rather their sense that their 'personality is entirely estranged from the sex to which he [*sic*] belongs'.[152] The patient grappled with their feelings, describing them as a long-held desire to be female: 'I can define my disposition no better than to say that I seem to be a female in a perfectly formed male body.'[153] The doctor, William Lee Howard, recognized this as an echo of Ulrichs's female soul in a male body but seemed convinced that the formulation was 'an original conception' with his subject.[154] Similarly, C. W. Allen invoked Ulrichs when he interpreted the case of Viola Estella Angell, who dressed and worked as a woman but who had male genitalia, as one of 'psycho-sexual hermaphroditism'.[155] The interest in this case, Allen wrote in 1897, was its rarity: 'Even in asylum practice it is rare that one finds the male [*sic*] possessed of the delusion that he [*sic*] is a female. The subject of this report appeared to be honestly convinced that nature had intended him [*sic*] for a female.'[156]

Austin Flint of the Cornell University Medical College published a retrospective study of a young person whom he had encountered in 1895 when that person had been arrested in women's clothing in Central Park in Manhattan. The article in the *New York Medical Journal* in 1911 contains photographs of the person dressed in female attire and then naked, to demonstrate 'masculine conformation'.[157] They said that they had worked as a domestic servant and preferred 'women's dress and occupations and usually dressed as a woman'. Flint, who had read Auguste Forel's *The Sexual Question* (1908), categorized his case as one of sexual inversion, referring directly to a section in Forel's book called 'Sexual Inversion or Homosexual Love'.[158] (Forel, for his part, was heavily influenced by Krafft-Ebing.)

The interpretation of transgender before transgender as homo-sexuality remained influential. Claude Hartland, who published their St Louis-based memoir as a faux sexological text, stressing their own effeminacy and womanly characteristics (their childhood love of dolls and later cross-dressing), saw this perversity, their 'perverted nature', their 'disease', their 'abnormal passion', as lying with their love of 'my own sex' (meaning of males);[159] 'He [sic] has the delicate, refined tastes of a woman, and what is worse, her sexual desires for men.'[160] Predictably this 1901 text was later published as a work of homosexual autobiogra-phy.[161] The case of Ralph Kerwineo (born Cora Anderson), another of Emily Skidmore's trans men, who lived with a woman as husband and wife, and then met another woman in a dance hall and married her, was mentioned by Kiernan in 1914 but as an example of 'invert marriages'.[162] The Oregon trans man Alan Hart, a medical novelist and doctor, who began life as Alberta Lucille Hart, appears as patient 'H' in the Portland psychiatrist J. Allen Gilbert's 1920 article 'Homo-Sexuality and Its Treatment', outlining the background to Hart's 'Transformation' in 1917.[163] From an early age, H thought of himself as a boy, liking the activities that boys did, and felt natural wearing masculine clothing. H had a series of relationships with women, with H in the masculine role according to the descriptions relayed by the medical commentator: 'H took the masculine rôle here so completely that I. C. often said going about with her was exactly like going with a man.'[164] H did not completely cross-dress, but rather wore a compromise hybrid of men's garments ('men's coats, collars, ties, tailored hats, English shoes') and a skirt.[165] He attended medical school as a woman, but after graduation persuaded Gilbert to perform a hysterectomy. He thereafter cut his hair, secured 'a complete male outfit', and, changing his name to Alan Hart, 'started as a male with a new hold on life'.[166] Moving around America, he practised medicine and wrote novels. He was married twice.

There are several significant aspects of this early lesbian/trans case. One was H's persistence in his masculinity. He resisted psychological treatment in the direction of any 'of the characteristics common to the female', as Gilbert expressed it; 'she [sic] absolutely refused to run any chances of losing her [sic] general masculine psychological character-istic in exchange for any benefit that might be derived from a proper orientation of herself [sic] as a female sociological unit in the social world of sex'.[167] Hence the clinician's reluctant decision to facilitate the process. Another noteworthy aspect of Hart's case is his determination to adapt to his 'true nature' in the manner described above, in both the outward sense (with 'male attire') and inwardly with surgery (the removal of 'her [*sic*] uterus').[168] Finally, it is important to note Gilbert's persistence with his diagnosis of 'abnormal inversion', homosexuality, in the face of his acknowledgement that Hart was 'from a sociological and psychological standpoint . . . a man'.[169]

One of the first American medical studies of transvestism per se was B. S. Talmey's 'Transvestism' (1914), five case studies of cross-dressing, which began by citing Hirschfeld and argued for transvestism as a separate category, while recognizing (*pace* Prosser) that it had hitherto (until Hirschfeld) been 'considered a mere symptom of homosexuality'.[170] These studies, which included one of Otto Spengler, discussed later, demonstrated that the urge to wear the clothing of the other sex went beyond mere dressing. 'The peculiar anomaly of the patient', he wrote of the then unnamed Spengler, 'is the desire to be a complete woman . . . he [*sic*] often wished to be castrated to be more like a woman. He [*sic*] longs for the female form.'[171] Another of Talmey's cases, who had lived in Dayton, Ohio, relatively undisturbed, described 'his' feelings: 'I have thought for some time, I was possessed with a female spirit, or a female soul inhabited my body.'[172]

So the concept of transvestism influenced American thinking. When medical experts from Rochester in New York State published a study in 1931 that included a case where someone assigned male at birth was raised and married as a woman, they called it 'Transvestism or Eonism' (they were aware of both Hirschfeld and Ellis).[173] Louis London, a well-read New York psychoanalyst, used the word 'transvestism' in the title of his 1933 case summary of the New York actuary who would come home from work, dress in female clothing, 'wait on his [*sic*] wife and wash the dishes', and then masturbate (seventeen masturbation fantasies were listed).[174] 'His [*sic*] greatest hope is to change to a woman, and this fantasy dates back to the time in his [*sic*] adolescence when he [*sic*] yearned to be a girl.'[175] London claimed a tale of success: after ninety sessions of analysis, the 'patient was completely cured of his [*sic*] perversion'.[176] If true, which is highly doubtful, it would have been a rare case of trans psychoanalytic reversal.

Psychiatrists in Chicago who published a study of a trans woman in 1944 referenced Ivan Bloch, Ellis, Krafft-Ebing, and Albert Moll – mostly the German editions of their work – but favoured Hirschfeld's 'terminology "transvestite"'.[177] Their study included their patient's self-statement, encompassing feelings of femininity from an early age, wearing female clothing, experiencing menstrual headaches, attraction to men, and desire for 'an operation'; 'I am a woman and as such I will live.'[178] Though they treated her as a 'male in female dress' – whose request that a physician provide a certification that 'he is a she' was an 'absurdity', and her request for surgery verging on 'mental deficiency' – the point is that she made such demands, 'asked to have an amputation of his [sic] genitalia and a vagina constructed'.[179]

The influence of the European experts lingered into the end of the period before transsexuality, oscillating between the sexological greats. D. O. Cauldwell derived the name transsexual, or rather 'psychopathia transexualis', from Krafft-Ebing's *Psychopathia Sexualis*.[180] Benjamin drew on Hirschfeld.[181] Gutheil engaged with Hirschfeld in his think-piece for American psychiatrists on the psychology of transsexuality and transgender, demonstrating that the sex researcher was still relevant in the 1950s.[182] Hirschfeld may also have had an advantage in that his 'Transvestism' essay, stressing the mixed sexual complexion of cross-dressing, in both its surface (transvestism) and deeper (transsexual) senses, was published in the USA in 1948 on the eve of the transsexual moment.[183]

Trans Moments

Transness surfaces at particular moments. In 1917, a doctor called R. W. Shufeldt published his notes on the Brooklyn fairy J. W., tantalizingly known as 'Loop-the-Loop', perhaps after the Coney Island roller-coaster but more likely referring to mutual oral sex.[184] Shufeldt, who was in the army when he published his study, had interviewed and photographed J. W. twice, once in 1906 (when his subject was in their early twenties) and again in 1916 (when they were in their early thirties). 'Few writers in the field of psychiatry have enjoyed what I had next', wrote Shufeldt, 'the opportunity to observe in the life incidents of this subject: the putting on of female attire by a contrary sexed male [*sic*]'.[185] And the medical observer watched and outlined the cross-dressing and application of powder, pomade, and 'fearful blond wig', while constantly mocking J. W.'s mannerisms and denigrating the tawdriness of his subject's efforts.[186]

Shufeldt saw J. W. as 'a typical example of contrary sexual instinct', a 'passive pederast'.[187] When cross-dressed, J. W. had paid sex with numerous men (between the thighs and in the anus). 'He' mentioned satisfying 'as many as forty men in twenty-four hours' and a line-up of twenty-three men in a Brooklyn room. One of the numerous 'husbands' and 'lovers', a musician, was interviewed, too.[188] But it was Loop-the-Loop's insouciance that surprised the medical expert, rather than the congresses with men:

> Lost to every sense of shame; believing himself [*sic*] designed by nature to play the very part he [*sic*] is playing in life, it was truly remarkable to hear this nervous, loquacious, foul-mouthed and foul-minded 'fairy' of the most degraded slums of a multi-millioned city chatter about his [*sic*] experiences, just as though he [*sic*] were talking about the rearing of fancy pigeons, or anything of a similar nature.[189]

The doctor presented J. W. as 'he' – indeed, stressed that 'he' was 'perfectly virile', referring to 'his' bodily attributes when naked – and provided photographic evidence of 'his' normal masculinity.[190] J. W.'s anus was just an anus; confirmed with a photograph of the posterior.[191] But the case details – references to pierced ears and earrings, menstrual claims, assertions and denials of pregnancy, an alleged misplaced pride in breast development, and the 'yearning' for female work and female attire – made it clear that this was no ordinary young man, whatever the sexual practices.[192] J. W. referred to 'his' anus as a vagina; hence the photograph to demonstrate otherwise.[193] Shufeldt classed J. W. in terms of homosexuality, but in later times the categorization may well have been transsexual.

Similarly with the memoirs of the fairy Jennie June, who operated in working-class New York in the 1890s and 1900s, fellating hundreds of young male migrant workers (Irish, Italian, Jewish) and soldiers and sailors.[194] 'Up to my early thirties', she wrote, as Ralph Werther, 'they always regarded me as a girl and used the feminine pronoun'.[195]

Werther's editor considered this a case of homosexuality or sexual inversion. But Jennie June was consistent in her belief that she was a 'girl without a vagina'.[196]

Werther / June / Earl Lind's *Autobiography of an Androgyne* (1918) is a remarkable text. It presents its author as both 'sexologist and sexological subject', sometimes with third-person distancing.[197] Werther or Lind (their other alias) was familiar with the sexological literature and consulted some of the American sexologists themselves, researched in the library of the New York Academy of Medicine, published under the imprint of the *Medico-Legal Journal*, and, like any self-respecting scientist of sex, rendered the more explicit descriptions in Latin. And yet much of the book is the sexual history of the fairy June, an explicit account of sex in the streets of Manhattan: the Bowery, Hell's Kitchen, and Mulberry Street were among her main haunts. Jennie's sexual partners referred to her as 'she', but these working-class consorts would have been unlikely to decipher the Latinate graphic detail of what they did to her or what she did to them.

In his learned introduction, the editor of the *Journal*, Alfred W. Herzog, situated Werther in terms of 'inversion' or 'homosexuality': 'a human being born with a body with sexual organs all those of the male, yet most likely with a body which shows certain earmarks of the female, and with a soul nearly all female, but certainly entirely female in regard to the sex question'.[198] 'Such a person', Herzog continued with logic now alien to modern trans sensibilities, 'is a homosexualist, because he [*sic*] feels like a woman and to him [*sic*] all male persons belong to the opposite sex'.[199] The term 'androgyne', for both author and commentator, captured Werther's gender ambiguity.

But if inversion was the sexologist's nosology, Werther was more considered in their classification. There are several places in the text where an earlier self-description of 'urning' (homosexual) is crossed out because it does not correspond to their reconsidered feeling of self.[200] They explained that they sometimes used the term hermaphrodite to their working-class contacts because they knew that they would understand the term even though they did not consider themself such – other than in a psychological sense. (The redacted term 'cocksucker' would have been the descriptor they would have favoured.) 'Androgyne', they felt, corresponded better to what they were, and quoted Krafft-Ebing:

'There is yet wanting a sufficient record of cases belonging to this inter-esting group of women in masculine attire with masculine genitals.'[201] They never doubted their femaleness: 'Your author is really a woman whom Nature disguised as a man.'[202] Werther elaborated, clearly coun-tering the trope of inversion:

> The girl-boy with diffused minor abnormality in physical structure, consisting in approach to the feminine type, is rather a female who has, along with some other male structures, developed testicles and penis in place of the usual ovaries and cunnus [pudenda]. Here it is not so much a case of a female brain in a male body, but of the female brain in a *female* body with various abnormal developments along the lines of male structure. A girl-boy is sometimes even physically perhaps more a female than a male, although the primary sexual determinants and some of the secondary sexual characters are those of the male sex.[203]

'I have been doomed to be a girl who must pass her earthly existence in a male body', they wrote when discussing their teenage years; 'I would meditate taking my father's razor and castrating myself in order to bring my physical form more in accord with that of the female sex to which I instinctively yearned to belong.'[204] Hence, when they were medically castrated, which they were when in their late twenties, Herzog recog-nized that it was to 'be more alike to that which he [*sic*] wished to be'.[205] *Autobiography of an Androgyne* deserves to be remembered more as an early American transgender text than as a work in homosexual history.

Hidden in Plain Sight

Just as the sexologists, many of them, treated transgender as a species of homosexuality, it seems likely that trans history has been submerged in the histories of same-sex sexuality. Even those sensitive to the blur-ring of heterosexuality and homosexuality and the multiplicity of queer cultures in pre-Stonewall America may not have been quite so alert to pre-trans configurations.

It is important to grasp the historical ubiquity of male effeminacy and female masculinity. Students are always amazed to learn of the popularity of female impersonation at the beginning of the twentieth

century, with the carefully crafted femininity of Julian Eltinge and
Bothwell Browne turning them into major vaudeville stars.[206] As Nan
Alamilla Boyd has pointed out, their success spawned a procession
of other impersonators, with, in the words of a contemporary critic,
'all the symptoms of homosexuality worn on their sleeves'.[207] Ray/Rae
Bourbon, whom we will encounter later as he was still performing in
drag in the 1950s, had a repertoire in the 1930s that included 'Since
Ivan Started Divin' – 'the girlies are arriving since Ivan started diving'
– one long reference to cunnilingus; and 'Gigolo', which told the story
of a man who left his wife for her boyfriend: 'My domestic bliss has
gone quite fluid . . . You can see this whole arrangement is just a wee
bit queer.'[208]

The fairy, usually discussed in terms of the history of male homosexuality, should be included in the story of transgender before transgender.
The Chicago pimp, Iceberg Slim (Robert Lee Maupin), wrote of 1938
that 'There's no such thing as a lady in our world. You either got to be
a bitch or a faggot in drag.'[209] Leo, an 18-year-old, African-American,
effeminate man in Chicago in the 1930s, thought in terms of fags,
faggots, cats ('colored queers'), bells, and queer people (on the one
hand) versus Jam people, straight people, squares, and those who were
manly (on the other). He felt that he could be himself when 'among
the faggots', and attended parties where 'almost all of the faggots there
were in drag'.[210]

David K. Johnson has outlined the 'remarkable degree of openness
and visibility' of what would later be termed trans women in Chicago
in the 1920s and 1930s; the University of Chicago's sociologists interviewed them in what have become vital historical sources.[211] Historians
have explored the pansy and lesbian crazes in the nightlife of the 1920s
and 1930s in New York, Chicago, and Los Angeles, a fascination with
effeminate male entertainers, female and male impersonators, and with
the fairies and masculine-presented lesbians who attended the various venues of night-time leisure: the tearooms, speakeasies, cabarets,
clubs, drag balls, and restaurants.[212] Gladys Bentley's Bobbie Minton
top hat and tails and Walter Winston's Gloria Swanston heels, skirt,
corset, and 'well-modelled bosom', were merely representative of a
wider phenomenon.[213] Thaddeus Russell has termed Harlem at this
time 'the most liberated public space in U.S. history' and contrasted

this openness with the 'black heteronormativity' encouraged by the later civil rights activists of the 1950s.[214]

 Pansies are part of the decor in the speakeasies and various clubs in Claude McKay's novel *Home to Harlem* (1928): 'The pansies stared and tightened their grip on their dandies. The dandies tightened their hold on themselves ... Dandies and pansies, chocolate, chestnut, coffee, ebony, cream, yellow, everybody was teased up to the high point of excitement.'[215] Even at the start of the novel, when the absent Jake yearns for the women of Harlem, their allure is framed in terms of the glamour of the fairy: 'Brown girls rouged and painted like dark pansies. Brown flesh draped in soft colorful clothes. Brown lips full and pouted for sweet kissing. Brown breasts throbbing with love.'[216] A. B. Christa Schwarz has commented on the 'inappropriateness of an unquali- fied transference of the concepts of hetero- and homosexuality into McKay's real and literary world'.[217] The lyrics to the African-American blues singer Ma Rainey's 'Sissy Blues' (Thomas Dorsey, 1928) capture the situation perfectly:

I dreamed last night I was far from harm
Woke up and found my man in a sissy's arms . . .

Some are young, some are old
My man says sissies got good jelly roll . . .

My man's got a sissy, his name is Miss Kate
He shook that thing like jelly on a plate . . .

Now all the people ask me why I'm all alone
A sissy shook that thing and took my man from home[218]

Her man had gone off with a fairy, but it was no more remarkable than leaving her for another woman, the more common theme of such music. It was her man's infidelity that was the subject of comment, not his sexual identity or masculinity. Women had to worry about competi- tion from fairies as well as other women.

 Black working-class culture was especially receptive to this gender and sexual flexibility. Kim Gallon has highlighted the role of the newspaper the *Baltimore Afro-American* in covering the 'gender-nonconforming'/'homosexual' expression of black female impersonators in Washington, DC and Baltimore during the interwar years.[219] Attitudes were not always affirming. Ralph Matthews referred to 'freaks of every nature ... The march of Gloria Swansons, Greta Garbos, Clara Bows, and Mae Wests goes on apace, dragging with it the morals of the impressionable youths who flock to these joints.'[220] Yet such reports were a confirmation of the presence of the 'freaks': it was, after all, a pansy *craze*. As Gallon has hinted, the *Baltimore Afro-American*'s reservations did not prevent it from giving full-page photographic spreads of leggy female impersonators headlining other attractive examples of black womanhood – nurses, students, trainee teachers, visitors from Ohio.[221]

Walt Lewis, an African-American hustler (sex worker) who provided a sexual history to those previously mentioned University of Chicago sociologists, talked of fags, 'women impersonators', and 'bulldaggers' (or women-fuckers) rather than homosexuals and lesbians, implying the significance of gender presentation rather than the sex of the desired sexual partner. 'Now the women fucker, she fucks women like men', he explained to the researcher; 'if she fucks your wife, or girl, your girl ... will leave you ... right away'.[222] Lewis was clear, and crude, about his enjoyment of sex with women – 'ways to fuck a girl' – but he was also highly matter-of-fact about sex with men:

> he had me to lay between his legs and roll it to him like a man does to a woman when he jazz her ... Some of them you cannot tell from a woman if they never have whiskers or mustash. They take in the ass, French you, like to be called girls names and if they like you will give you money, let you stay with them like a man and wife.[223]

In his unpublished history of hustling (male sex work), Thomas Painter said that in New York in the 1930s the 'flaming homosexual' was 'a common sight':

> there were blatant, painted, 'flaming' young homosexuals (older ones seldom try to flame) with marcelled hair quite long and held in place with innumerable hair pins, with mascaraed [*sic*] eyelashes, powdered

cheeks, lipsticked mouths, and riotously colored fingernails, their fingers vulgarly ornamented with large rings, their clothing femininely soft and loudly colored, their shoes suede, their whole appearance and manner calculated to attract the attention of normal men. These were to be seen any night in Times Square.[224]

Of course, there is a difference between effeminate presentation, with the feminine accoutrements described by Painter, and actual cross-dressing – donning a woman's dress. The fairy Jennie June adopted exaggerated female mannerisms – what she termed 'female-impersonation' – rather than actually wearing female clothing.[225] But Painter argued that many of the fairies and 'flaming queens' moved with ease into the numerous drag clubs as 'female impersonators'. 'This profession offers much better pay than street prostitution', Painter explained, and, apart from promiscuous sex and the drink, 'gives the young homosexual an opportunity to indulge to his full his longing to dress and act as a woman – or even more womanishly'.[226] (In New York, he added, such venues also 'feature Lesbian entertainers', presumably parading a form of female masculinity.[227]) But our observer did not really distinguish between sex work and drag artistry. The 'drag entertainer' was merely a more finessed hustler, 'holding out for a large number of presents or sums of money rather than doing a cash-on-delivery business'. They usually had a 'husband', 'some truck-driver or prize-fighter . . . largely kept by his homosexual "wife"'.[228] Though he sometimes referred to 'herself', 'her', and 'she' when discussing these entertainers, usually in parentheses, perhaps to reflect street argot, or possibly to indicate that they were not really female, Painter categorized them as homosexual: 'transvestist male homosexuals'.[229] See Illustration 4. As I have observed elsewhere, effeminacy and cross-dressing – the fairy and transvestite – are somewhat blurred in Painter's history of the 1930s.[230]

There is a case to be made, then, for reworking the history of the fairy specifically, and male effeminacy and female masculinity more generally, in terms of transgender before transgender. Boyd negotiated this successfully in her 2003 study of San Francisco, writing of a 'public culture of homosexuality and gender transgression' based on the gender expressions of club and stage. She stresses the complexities of identities and desires, the sexual dialectic (my term, not hers) between

4   *A cross-dressed owner of a male brothel, 1930s/1940s, a friend of*
Thomas Painter.

stage and street and between audience and performer during the first half of the twentieth century.[231]

Certainly, the more I view Paul Cadmus's New York paintings *Shore Leave* (1933), *Sailors and Floosies* (1938), and *The Fleet's In!* (1934), the more I am convinced that several of the 'floosies' were male-to-female cross-dressers. The centrally placed woman in *The Fleet's In!* has rather boyish breasts and a prominent Adam's apple.[232] Here – though none of the experts seem to have noticed it – is a trans woman, hidden in plain sight.

Language of the Streets

The psychiatrist George Henry's *Sex Variants* (1941) also contains case studies of what might later be termed transgender individuals, though any detected imbalances in masculinity in women and femininity in men were again classified in terms of homosexuality. One was the pseudonymous Victor R., aged 23, with a 'distinctly feminine appearance'.[233] At the time of their interview, Victor was dressing in women's clothing at home and did all the housekeeping in the apartment, providing what they called 'a woman's touch'.[234] They said that they wished they were 'a girl' but described themself as 'queer'.[235]

Antonio L., another of Henry's informants, was a somewhat incongruous figure, with the 'build of a prize fighter', and surviving through paid sex with other men, but who, by the time they were 30, had come to feel woman-like in a very transgender way, and in terms of both sexuality and gender: 'I would like to be a girl . . . I like to wear girl's clothes and at home I regularly wear complete female attire. In sex I act as a woman. I want to be the woman.'[236] Daniel O'L., Antonio's friend, was an effeminate hustler who had exchanged life on the streets for a more permanent relationship with a man who treated Daniel like a wife: 'He likes to see me dressed as a woman';[237] 'When we go to parties I dress just like any other young girl. I wear a stuffed brassiere and a switch. I use powder and I paint my fingernails and also my toenails if I wear sandals. I like strong perfumes and I use them at parties. The men like it also.'[238] Although Henry predictably summarized this case as one of homosexuality as 'an expression of innate effeminacy', Daniel's (mediated) self-analysis was different, and again very trans-like: 'I've always wished I was a girl . . . I've always thought I would like to have the hair taken out of my face. I would like to be castrated and made a girl now if it was possible.'[239] So there were people in the era before transsexuality and transgender who exhibited the characteristics later attributed to transness.

There were transvestites, too. Moses I., who was wearing stockings, a corset and silk underwear under their clothing when being interviewed by Henry, claimed their cross-dressing was inspired by seeing a performance by the famous female impersonator Julian Eltinge; 'It made me feel I was sinking into something, that I was giving up something.'[240] But Moses's feelings went deeper than wearing women's clothing. They contemplated castration: 'They could operate on my pituitary gland and give me ovarian extract. In time I could assume the role of woman and follow the inclinations I have bottled up for so long in my heart of hearts – to do nursing, looking after children.'[241] Moses described impulses to self-mutilate: 'I would prefer to have my penis removed and a vagina made surgically but sometimes at night I get the idea of doing it myself.'[242] They also said that they 'had a desire to be a woman with a man'.[243]

Rudolph von H., another of Henry's transvestites, was actually Otto Spengler, who knew both Benjamin and Hirschfeld and was one of the subjects in one of the first US medical studies of transvestism, published in 1914, with pictures of them cross-dressed and then nude as Psyche as in Frederick Leighton's famous painting *The Bath of Psyche* (1890).[244] When Hirschfeld visited the USA, Spengler attended the lectures and presented as a 'typical transvestite'.[245] They were in their sixties at the time of the interview and in straitened circumstances. Spengler had long hair, 'fairly well-developed breasts', pierced ears, used cosmetics and perfume, had been cross-dressing since an early age, and had a 'wardrobe of seventy to a hundred dresses' (Henry commented that they were rather trapped in the fashion era of their younger days).[246] Spengler admitted that women's clothing 'was a sort of fetishism' initially, and they would masturbate while cross-dressed but implied that it later reflected their feeling that they were a woman; when they had sex with their wife, they thought of themself 'as a woman'.[247] They had been taking progynon, an early form of estrogen, to develop their breasts: presumably they are the subject of the photographs in an Appendix to Henry's study.[248] Spengler had undergone the Steinach vasectomy operation for rejuvenation (probably performed by Benjamin for he was using the technique in the USA and Spengler was a patient of his), and had had their testicles X-rayed to induce sterility – all of which indicate that such gender-modifying techniques were available in America in the inter-war period, given that Spengler had been in the country since the age of 19.[249] It is significant, in the light of the earlier discussion of Hirschfeld's 'heterosexual' transvestites, that Spengler claimed that the sex researcher had told them that they (Spengler) were an 'inverted Lesbian, that I would attract men but would rather have women fondle me'.[250]

It seems clear that there was a difference between the languages of the streets and the classifications of academics and medical and psychiatric experts. The 1930s studies of both Henry in New York and the sociologists in Chicago attribute homosexuality to sexual configurations that in day-to-day interaction were consistently described in other ways. We may never know what the original transcripts said in the New York study; we have only the printed version. But the Chicago study provides access not only to the initial statements but also to the sociologists' attribution of meaning to their findings. Letters collected by the interviewers and the life histories themselves all referred to a queer effeminacy rather than homosexuality per se: 'queer', 'Swishy bells', 'cock suckers', 'pansy', 'Nellie', 'Miss Millard', 'fellows that wore make up and had plucked eyebrows', '"Showy" people', 'Margie queen'.[251] Even the jokes collected invariably refer to queers, queens, and cocksuckers rather than homosexuals.[252] Yet in the actual glossary, words that may not really have connoted the precise identity 'homosexual' were marked as such. Thus, a 'belle' was 'any homosexual person'; 'Fairy-joint' was a 'place where homosexuals gather or can be found'; and, most tellingly, 'Queer' became 'anything or anyone homosexual'.[253] My point here is that the sexual and gender imprecision of the streets may have been more attuned to pre-trans identities than has previously been acknowledged.

Sex Reassignment Before Sex Reassignment

Though it was a time before developments in surgery and endocrinology facilitated gender modification, we have seen that several people availed themselves of the techniques accessible: castration, sterilization, removal of the ovaries, hysterectomy, mastectomy, even vaginoplasty, and electrolytic depilation, facial surgery, and hormone treatment. Those who did not have such access longed for these interventions. Indeed, given the experimentation of Eugen Steinach, Benjamin, and others with endocrine surgery in the 1920s and 1930s, the transplanting of gonads in animals and vasoligation and adrenalectomy in humans, and the impact of glandular research and synthetic hormone experimentation on wider literary and visual culture, it was inevitable that the quest of changing sexual characteristics and rejuvenating mental and physical properties would be reimagined as gender reassignment.[254]

The female impersonator Ray Bourbon had a 1941 song, 'Gland Opera', which extolled the rejuvenating properties of glands:

Now a certain worn-out bachelor, who had gone from frail to frailer,
They had no stevedore's glands in stock,
So they gave him a ladies tailor's
But now he's camping at the beaches, making passes at sailors.[255]

Gladys Bentley, whom we have also encountered in the pages above, proclaimed in 1952, on the very eve of Christine Jorgensen's media moment, that female hormone treatment had saved her from 'that half-shadow no-man's land which exists between the boundaries of the two sexes . . . I became a woman again.' She moved from lesbian icon of female masculinity to a 'happily married' woman 'living a normal existence'.[256] It would not have required much imagination to invoke the possibilities of gender shaping. Joanne Meyerowitz has referenced a trickle of enquiries to the advice columns of the American popular magazine *Sexology* in the 1930s from people who had read or heard about 'change of sex' and who were seeking further information: 'I have a peculiar complex – I believe it is called "Eonism". That is, I desire to dress as a woman . . . The fact is I have an even stronger desire, and that is – I wish I were a woman . . . I am interested in the Steinach operation in regard to change of sex. I would like more information.'[257]

Outside the USA, individuals had experimented with assuaging their gender unease. In England, Roberta Cowell, who would now be considered a trans woman, self-dosed oestrogen in the 1940s. Michael Dillon, a trans man, an Irish medical student, travelled between Ireland and England during his sexual confirmation surgery, carried out also in the 1940s by Harold Gillies who specialized in reconstructive surgery, mainly for those suffering the ravages of war.[258] Dillon, both knowledgeable and resourceful, had begun taking testosterone in the late 1930s; his biographer Pagan Kennedy claims it as the first recorded use of the drug for the purpose of gender reassignment.[259] Dillon had a double mastectomy, and his constructed penis, the result of seventeen operations, must also rank as a pioneering example of such surgery.[260] Furthermore, Dillon's book *Self* (1946), though it did not use the actual words 'transsexual' or 'transsexuality', was conversant with the effects of male and female hormones and wrote about bodies being reshaped to fit minds years before the case of Jorgensen and the work of Benjamin began to popularize such concepts.[261] Dillon referred to those who were members of 'the sex to which he or she feels he belongs, but to which he is not recognized as belonging', and advocated the injection of hormones to achieve a 'tolerably happy life for the individual concerned'.[262] 'Surely, where the mind cannot be made to fit the body, the body should be made to fit, approximately, at any rate, to the mind, despite the prejudices of those who have not suffered these things, yet to suffer which they so readily condemn others.'[263] (Though it should be noted that Dillon saw such cases as a type of homosexuality.[264]) Moreover, he performed a secret orchidectomy (removal of the testes) on Roberta Cowell in 1950 or 1951.[265]

Harold Gillies would go on, in the early 1950s, to perform a vaginoplasty on Cowell as well as extensive cosmetic surgery – though by then this was post Jorgensen and Benjamin.[266] Alongside very detailed photographs of penile and vaginal reconstructions, Gillies wrote almost nonchalantly of the latter: 'Quite simply – the body and shaft of the penis were slipped out of the loose hairless skin cover, the penis was discarded, and the skin envelope closed at its free end and invaginated into a prepared cavity. It was like a finger-stall and indeed was pushed into place on the surgeon's finger.'[267] Also in England, Gillies's colleague, the surgeon Archibald McIndoe, constructed over sixty vaginas (in women) from the late 1930s until 1950, so the surgical techniques existed.[268]

In Switzerland in the 1930s and 1940s a man had his penis and testicles removed, though the precise status of his transsexuality is unclear from the doctor's account.[269] In the early 1940s, Arnold Leber, assigned male at birth but convinced of her femaleness, had hormonal treatment, electrolysis, and genital surgery to confirm her sense of identity, becoming Arlette Leber.[270] Leber's surgeon thought that the provision of 'an artificial vagina' was 'a useless luxury'.[271] But the point is that (in the words of the court that confirmed Arlette's female status) he performed the procedure to make Leber 'anatomically what he [sic] is already psychically and what he [sic] wishes so much to be: a woman'.[272] And all this was pre-Jorgensen.

Then there was Lili Elbe, born as Einar Wegener, the Danish artist known as Andreas Sparre in the book outlining this case, who underwent a series of treatments in Germany in 1930 in an effort to reconcile her male body to her increasingly felt female self. Hirschfeld was one of those she consulted. Her rationalizations varied, seeing herself first as a being of two sexes, male and female, and then increasingly viewing her male bodily attributes as an alien presence: Lili had to remove Einar in order to survive. The precise nature of the surgeries which were carried out before her death in 1931 are rather opaque but included removal of the male sexual organs and the transplantation of ovaries from a female donor. The ovary transplanting has shades of Steinach, and Kadji Amin is surely correct in positioning Elbe's case as one of Steinachian rejuvenation and glandular psychology rather than an early example of gender transition and wrong-body narrative.[273]

There is also the issue of intersex to consider – those who were once called hermaphrodites: people with ambiguous bodies whose bodily uncertainty came to be resolved through recourse to surgery. There is a sense in which, in an era before effective surgery and endocrinology, cross-dressing, and assuming the name and occupations of the opposite sex, were the equivalent of sex reassignment; Geertje Mak equates them in her history of intersex.[274] So there is some overlap with the discussion on cross-dressing and trans-ing. From the 1890s onwards (roughly), anaesthesia and surgery were increasingly employed to initiate more far-reaching bodily modification: clitoridectomy (removal of the clitoris), hypospadias repair (of a penis with a misplaced urethra), the removal of breasts, vaginal construction or reconstruction, penectomies (amputation of the penis), orchiectomy or orchidectomy (the removal of one or both of the testicles), and castration (the removal of both testes).[275] In one notable New York case in 1903 – controversial both because the patient was asked for their preferred reassignment and for the reason that the surgeon may have mistaken their dominant sex – the clitoris was removed and its skin used to enlarge the vagina. Curiously, in the paper's subtitle the clitoris was named as a penis: 'OPERATION FOR REMOVAL OF THE PENIS AND THE UTILIZATION OF THE SKIN COVERING IT FOR THE FORMATION OF A VAGINAL CANAL'.[276] It could almost have been the heading for an article on male-to-female transsexual surgery in the 1960s. In another intriguing parallel to the history of transgender, Christina Matta and Elizabeth Reis have shown that a crucial motivation of intersex surgeons was an unease with homosexuality and a preoccupation with heterosexual (married) intercourse: early intersex and transsexual surgery shared a heteronormative impulse.[277]

An article in the medical journal *Surgical Clinics of North America* set out some of the possibilities in 1929 when discussing the situation of a young woman with what was considered to be a large clitoris: 'Shall we attempt to reconstruct the urethra, obliterate the vagina, attempt to make a scrotum and liberate the penis, or shall we amputate the penis and remove the testicles?'[278] Or in Baltimore in 1935, at the Johns Hopkins Hospital, Hugh Hampton Young held out the possibilities of transformation to a black patient Emma T: 'Would you like to be made into a man?' Young assured her that it 'would be quite easy, that it would only be necessary to remove the vagina, and do a few plastics to carry the urethra to the end of the penis'. But Emma decided to keep both her penis and vagina; the former afforded her pleasure with her various women friends, and the latter was her 'meal ticket' to support by her husband.[279]

The gender dysphoric could draw on the possibilities exhibited by the treatment of the intersexed. Some form of gender reassignment seemed entirely possible before the transsexual moment.

The history of gender reconciliation is not limited to the post 1950s and 1960s, then. To quote the always perceptive Meyerowitz, 'the concepts of sex change and sex-change surgery existed well before the word *transsexual* entered the medical parlance'.[280] The more famous examples of modification – Elbe, Dillon, Cowell, and, in Britain in 1936, the case of the Olympic athlete Mark Weston – were renowned because they were publicized in the press.[281] In the words of David Andrew Griffiths, 'Developments in biomedical science and stories in the press provided a language with which individuals could approach the medical profession to request surgeries to enable their bodies to better conform to their inner sense of sex.'[282] Julian Gill-Peterson, who has accessed the archives of Johns Hopkins for the 1930s, has uncovered a series of examples of precisely those sort of people, travelling to Baltimore to ask the intersex experts to alter 'their external organs' to match their 'personality'.[283] In an era of what Gill-Peterson has described as 'an entangled field of inversion, hermaphroditism, homosexuality, and transvestism', people did muse about their bodily configurations and the possibility of changing (or reconciling) them.[284]

Conclusion

I have attempted to make sense of what Meyerowitz has described as 'the many "inverts", "androgynes", "men-women", and "masquerades" of the early twentieth-century transatlantic world'.[285] We have seen the makings of transgender in the various presentations of self that often strained against the interpretive frameworks of those who sought to classify those who cross-dressed or who felt a gender disjunction far deeper than mere clothing. We have encountered 'men' who felt exactly like women, and 'women' who felt like men, penises that were clitorises, and clitorises that were penises, masculine souls heaving in feminine bosoms and feminine souls in the bodies of 'men', those assigned female at birth who felt that they were essentially men, those who felt that wearing the clothing of the 'opposite' sex reflected their inner self, 'men' who referred to one another as girls and adopted female names, women without vaginas and men with vaginas, women with penises, 'women' who fucked like men, 'men' who kept house (and more) for their husbands and 'female husbands' (assigned female at

birth but living as men and married to women), 'women' who wanted
to live as men and 'men' who wanted to live as women, women in
masculine attire with masculine genitals, men who merely wanted to
dress as women or wear some form of women's clothing (often with
erotic motivation), and those who were more committed to presenting
or living as the 'other' sex or who actually felt/knew that they were
members of that sex.

Numerous cases of female masculinity and male effeminacy were
once too easily classified as types of homosexuality, by both sexologists
and historians. Now, without doing injustice to the discipline of the
historical context, or forcing a fully formed transsexuality on disparate
formations, they might better be seen as prefigurements of transgender: trans before trans.

2

The Transsexual Moment

Introduction

The invention of the term 'transsexual' is often attributed to Harry Benjamin, an early expert and therapist on such matters, who used the word 'transsexualism' in a lecture in 1953, held before the Association for the Advancement of Psychotherapy.[1] It was in a symposium on 'transsexualism and transvestism', in a lecture called 'Transsexualism and Transvestism as Psycho-Somatic and Somatic-Psychic Syndromes', in which Benjamin not only used the word but set out the definitions that would distinguish between cross-dressing and bodily modification. The transvestite wants to play the role of the 'opposite' sex: 'The male transvestite admires the female form and manners and tries to imitate both with an intensity that varies greatly from case to case.'[2] Transsexualism is of a different order: 'It denotes the intense and often obsessive desire to change the entire sexual status including the anatomical structure. While the male transvestite *enacts* the role of a woman, the transsexualist wants to *be* one and *function* as one, wishing to assume as many of her characteristics as possible, physical, mental and sexual.'[3] Benjamin was convinced that the difference between the male transvestite and the transsexual lay with their attitude to their sex organs: 'In transvestism the sex organs are sources of pleasure; in transsexualism they are sources of disgust.'[4] This chapter examines the early history of transsexuality from the time of its naming, and the role of transvestism (cross-dressing) in this supposed transsexual moment.

Transsexuality

John Money credited Benjamin's *The Transsexual Phenomenon* (1966) with introducing the word transsexual into the vernacular.[5] However, a version of the name was actually first used in print in 1949 by D. O. Cauldwell in the popular magazine *Sexology*. The term was 'psychopathia transexualis', in obvious reference to Richard von Krafft-Ebing's famous nineteenth-century sexological text, *Psychopathia Sexualis* (1896), 'a pathologic-morbid desire to be a full member of the opposite sex'.[6] Then, in 1950, and a little less reprovingly, Cauldwell wrote of 'transsexuals', 'individuals who are physically of one sex and apparently psychologically of the opposite sex'.[7]

But, even then, the emphasis was on transvestism. Christian Hamburger and his Danish colleagues, of Christine Jorgensen fame, used 'genuine transvestism' and 'eonism' (Havelock Ellis's term) to describe those who required (demanded) surgery to reconcile their bodies with their sense of self – and this was in 1953.[8] Most of *Sexology*'s articles in the early 1950s, anthologized in Cauldwell's edited collection *Transvestism . . . Men in Female Dress* (1956), including a piece by Benjamin called 'Trans-sexualism and Transvestism', were indeed about transvestites rather than transsexuals.[9] One of *Sexology*'s rare pieces dealing with female-to-male transsexuals (and not in the Cauldwell volume) was by the British/Australian sexologist Norman Haire, who resorted once more to transvestism to describe a patient treated with implanted pellets of male hormone, describing them as a 'transvestist'.[10] There were also more letters from transvestites than potential transsexuals sent to the editors of *Sexology* in the 1950s, which, in turn, reflected the attitude of the magazine: supportive of cross-dressing but less sympathetic to bodily transformation.[11]

So the actual history of US transsexuality is relatively short, and, as already intimated, we could locate its effective beginnings in the 1950s, despite an earlier history of desired bodily adjustment that has

5   *Publicity for* Glen or Glenda?

been discussed in the preceding chapter. Joanne Meyerowitz began her story with the publicity surrounding the sex change of Christine Jorgensen, a media event of epic proportions – similar to the impact of Alfred Kinsey's research on sex in the 1940s and 1950s – that brought the possibility of what was then called 'sex change' firmly into the public sphere.[12] Meyerowitz has written of a 'fierce and demanding drive' for bodily modification during that decade.[13] Ed Wood's famous movie, *Glen or Glenda?* (1953), dates from this 1950s obsession with the possibilities of surgery. See Illustration 5. Jorgensen's Danish specialist received hundreds of letters from men and women – from Algiers to New Zealand – who wanted to transform their sex.[14] American transsexuals, predominantly male-to-female in these earlier days, sought hormonal treatment from Harry Benjamin in New York, and surgery from approachable practitioners such as Elmer Belt in Los Angeles and the Langley Porter Medical Clinic at the University of California School of Medicine in San Francisco.[15] Or they travelled to Casablanca for treatment by Georges Burou, pioneer in pedicled skin inversion vaginoplasty (the conversion of penis into vagina).[16] The first editions of the trans magazine *Turnabout*, which appeared in 1963, have a couple of knowing references to trips to Casablanca.[17] *Sexology* (in 1966) was not exactly encouraging about the availability of such treatment, but was clearly aware of it:

> The best known place in the world today where such a sex change operation could be done is by one doctor in North Africa. The cost runs into many thousands of dollars. You would have the cost of transportation to North Africa, you have the cost of the stay in the hospital, and you have the cost of the surgeon's fee, the last two of which amount at present to three or four thousand dollars.[18]

But we are getting ahead of ourselves.

To gauge the pace of change, we could consider the summary of rudimentary knowledge in San Francisco in the late 1950s in a report co-authored by the Medical Superintendent of the Langley Porter Medical Clinic. The authors, Karl Bowman and Bernice Engle, observed the desire of some patients for the surgical transformation of their bodies, and the reluctance of many surgeons to perform such reconstructions, and suggested that the Clinic refused rather than granted permission for such operations: as psychiatrists, they favoured psychotherapy rather than surgery in any case, given that those treated were classified as 'sexually deviant individuals'.[19] Interestingly, they included notes on a married couple who remained together (though divorced) after changing roles. The 'man', although refused castration and a penectomy, took female hormones, assumed the female role in the household (that is, kept house) and took a female name. The 'woman' took testosterone, had breast reduction, 'took a man's name, dressed as a man, and worked outside the house as a man'.[20] Significantly, in the light of earlier discussion, Bowman and Engle termed such cases transvestism rather than transsexuality.

There was wariness in the 1950s, then, towards surgical solutions. Cauldwell, the man who named transsexuality, was fending off enquiries about removal of genitals and the creation of vaginas during the late 1940s and early 1950s. He was highly discouraging. As one correspondent put it, 'You told me that the chances were that a surgeon who would amputate my healthy genitals for a price might decapitate me as willingly, if either was to be done in consideration of a fee.'[21] In 1950, when W. S. from Illinois asked, 'My greatest desire in life is to become a woman and I would like to know how I can secure the help of surgeons in the physical metamorphosis into a woman', Cauldwell replied, 'you know it takes a great deal more than surgery to turn the male into female. A serious problem involving biology, psychology and physiology would be involved. Although surgery can create and establish a vaginal canal in the male, it can not create the nerve responses that would give a true sense of femaleness'; and he lectured W. S. on the duties of the surgeon to preserve healthy organs and of the legal risks involved.[22] Generally, if *Sexology* is any guide to popular advice (and the transvestite publication *Turnabout* sometimes drew on its expertise), its editorial counsel was not in favour of surgery in the 1950s: 'The truth is that Jorgensen was a male and has remained a male, although a mutilated one'; 'Change of sex is impossible.'[23] The magazine's attitude only changed (cautiously) in the 1960s, as reflected in the earlier quote about Casablanca.

A 1955 report in the *Journal of the American Medical Association* showed little sympathy either to those who sought sex transformation or to the potential surgery: 'It would appear that the operations have been performed more because of the desperate, pitiful state of the patients than on the basis of facts about the disorder.'[24] The authors, Los Angeles psychiatrists (they practised at the School of Medicine, University of California at Los Angeles), claimed that their subjects had 'extremely shallow, immature, and grossly distorted' notions of femininity (they were trans women), and were selective and restricted in their memories of their earlier lives, 'highlighting those memories that supported the concept of their having been female from birth' and producing histories that had a 'remarkable similarity'.[25] They were also often severely sexually conflicted: 'For some of them this took the form of a rather spectacular prudery, while for others the fear and guilt were more prominent.'[26] Consequently, the authors were not exactly supportive of surgery – indeed, they were wary of their patients' sense of urgency. They concluded that the severity and complexity of this psychological problem made it 'unlikely that it can be removed by amputation of the genitals'.[27]

The famous sex researcher Alfred Kinsey, though understanding of transsexuals – and, indeed, a pioneering collector of their sexual histories in the 1940s and 1950s – was also against surgery. He corresponded with Benjamin. He maintained contact with Louise Lawrence, a trans woman, who provided him with an autobiography, clippings, transcriptions of transvestite erotica, and an introduction to the worlds of cross-dressing and transsexuality.[28] The Kinsey Institute has become an important trans archive. See Illustration 6. Yet Kinsey's tolerance did not extend to trans surgery. As Meyerowitz has put it, 'It was here that he hit the limits of his sexual liberalism.'[29]

Early Days

Nonetheless, American trans people did seek out surgery – in varied stages of complexity. Benjamin's correspondence, now in the Kinsey Institute, indicates the problems involved in obtaining the desired medical intervention in the USA in the late 1940s. There were surgeons willing to operate, but legal opinion indicated that such surgery would be treated as 'mayhem'. The Attorneys General of both Wisconsin and San Francisco gave such advice.[30] A 1948 case summary from the Department of Neuro-Psychiatry at the State of Wisconsin General Hospital shows both the difficulty in obtaining treatment and the experimental nature of such medical knowledge. There was a meeting of staff to discuss the case, at which thirty people were present, and they recommended 'surgery (castration) and plastic surgery'. Interestingly, they had also contemplated 'brain surgery . . . that might destroy his present desire to remain a female mentally', but the patient was opposed to this proposed resolution and there was a risk that the end result might have been incapacitation. After all that, the Wisconsin Attorney General intervened and the Hospital was not permitted to perform the operation.[31]

Benjamin, whose patient this was, was quietly reassuring, writing that he would 'consider you definitely a woman that accidentally possesses the body of a man', and suggested X-ray castration and X-ray treatment

6   *A group of transsexuals, trans women, in the 1950s, part of the Harry Benjamin archive in the Kinsey Institute.*

of her face to remove hair growth.[32] The case did not end there. It is clear that Benjamin, possibly prompted by the patient's insistence, was trying to find a surgeon willing to operate – or, perhaps more accurately, a State that would allow the operation.[33] Yet by 1950 there were further case notes indicating a less favourable diagnosis. Kinsey and Bowman had been consulted and they agreed that surgery was not advisable, stressing the continuities rather than the changes brought about by medical intervention. Secondary sexual characteristics would remain. Sexual desire would continue, though without genital outlet. The psychological problem would persist. The patient would be a eunuch rather than feminized. The report advised alternative treatment: 'The patient should be encouraged to undertake homosexual relations as a means of learning to value his [sic] genitals.' Therapy should be aimed at 'reality reinforcement': 'He [sic] does *not* look like a woman; he [sic] can never be a true woman; he [sic] exaggerates others' interest and fondness for him [sic]; he [sic] can do something for himself [sic]. This must go very slowly as he [sic] is easily pushed to the edge of his [sic] defenses, becomes incoherent, agitated, and presents a near-psychotic picture.'[34] But the patient persisted. A 1953 letter from a medical specialist in Copenhagen indicated that she had had surgery (presumably the removal of the penis) and that a vagina could be constructed later.[35]

One patient had their penis and testicles (not the scrotum) removed in Holland. There was no vaginal construction at that stage, but they then went to Denmark where the scrotum was fashioned into labia. The idea was that they would revisit the vaginal construction later.[36] By the 1950s, then, some of Benjamin's patients were travelling to Europe and elsewhere for operative treatment. His files on more than 300 patients in the period up to 1966 show that Amsterdam and Mexico City were destinations in the 1950s, with Naples and Casablanca in the 1960s; 20 of his patients visited Burou in North Africa.[37]

We are fortunate that, in the 1950s and 1960s, Benjamin corresponded with a pioneer in transsexual surgery, the Los Angeles-based Elmer Belt, with whom he discussed both surgery (Belt's specialism) and hormonal treatment (Benjamin's area of expertise). The correspondence provides a unique – though medically focused – glimpse at this period in the nation's trans history. As Benjamin wrote to his friend in 1962: 'I realize more and more, Elmer, that we are engaged in an

7   *A pair of transsexuals, trans women, in the 1960s, part of the Harry Benjamin archive in the Kinsey Institute. Benjamin both photographed his transsexual patients and was sent photographs by them.*

entirely new type of practice, a truly untrodden field of medicine, and this realization often frightens me, but it is at the same time of course most intriguing, and most fascinating.'[38] See Illustration 7.

It is important to recognize that Benjamin and Belt were essentially sympathetic to their charges, usually used trans-appropriate pronouns (not always the case with therapists), and were saddened by the hardships that their patients endured. Benjamin was accepting of trans sex work: 'That she made her living as a prostitute does not lower her at all in my estimation. I can understand that, aside from economic factors, the profession appealed to her because it constantly reaffirmed to her inherent femininity.'[39] He understood drug addiction: 'I have seen it unfortunately happen in several cases of these transsexual patients who cannot get any help in this country.'[40] He tolerated patients who ignored his advice and went elsewhere for inadequate treatment or who administered their own surgery. There are references to self-laceration, bad breast implants, and the need for vaginal repair.[41] 'Yes, I have known her for quite a number of years', he reported of one, 'first as a boy, then after an operation without my consent (I believe in Mexico), and then through two marriages. I also know her third husband . . . I advised him . . . not to rush into marrying . . . but he did anyway. C'est l'amour.'[42]

While they were understanding and compassionate, they also felt safe with one another in complaining, as insiders, about the trials of patient interaction. 'You're quite right about the characteristics of those T.V.'s [sic]', Benjamin wrote to Belt in 1958; 'They sometimes seem to combine the worst traits of both sexes.'[43] In 1959, Belt referred to unpaid bills reflecting badly on future patients: 'I cannot take any transvestite into the Hollywood Hospital until the . . . account has been settled.' He continued, 'I believe, of course, that these people are picked on but, my gosh, they certainly are a screwy lot and they also have a yen for publicity which is almost an obsession.'[44]

In 1961, they discussed Patricia Morgan (formerly Henry Glavocich), whose autobiography reveals that she hustled to earn the money for her gender-affirming surgery.[45] Benjamin joked, 'I had given her strict advice to stay out of gay bars and similar places, but apparently she thought my advice did not include the West Coast.'[46] Belt reported that Morgan was 'arrested after using the lavatory for females in one of Hollywood's restaurants of very questionable reputation'. Another patient, who was there, told him that Morgan recognized someone and 'called out: "You hoo! I'm a woman now; I have been operated on by Doctor Elmer Belt. You hoo!" Then she used the toilet and got arrested.'[47]

Belt and Benjamin conversed about the efficacy of the available treatment. In 1960, Belt outlined a case of potential female-to-male surgery:

Tommy is a mannish, swaggering, deep voiced, somewhat hippy female, who at first glance looks every inch a man. Inquiry reveals the fact that there are four or five inches lacking in one area and quite a little too much in others. Tommy would like, first, to have the somewhat ample breasts amputated so that he can swim with less embarrassment. All of his feminine features are a great source of embarrassment to him, and well they should be, because he is bald and looks so mannish.[48]

Belt clarified that it was easy to justify a breast amputation but that the penis was a different matter:

To make a penis for Tommy I shall first have to make a 'suitcase handle' on the abdominal wall. Plastic surgeons call this a tube graft. The graft is then walked down to the pubis and finally planted just above the clitoris. Then the remaining abdominal end is freed, its end is closed and it is allowed to hang down between the thighs. In this last step a section of cartilage is removed from the last rib and inserted into the tube graft so that it is a rigid tube, made rigid by the cartilage. Since it is hinged on the pubic ramus it can be elevated and inserted into the introitus without any great difficulty.

Belt continued that, while this had been done for males after industrial accidents, he was convinced that if he tried to perform the same procedure for Tommy he would face opposition and was not sure what to do.[49] Benjamin confirmed that when he was in London the surgeon Sir Harold Gillies had shown him pictures of an artificial penis he had constructed, but agreed with Belt that there would be problems with 'the puritanical hospital boards'.[50]

They monitored the patients of the famous Georges Burou in the early 1960s. Belt expressed puzzlement at the use of penile skin for the vagina, a technique being perfected by the Frenchman. Belt claimed to have tried it in his earlier surgery and was candid about its reported failure: 'All of these early cases stenosed in no time at all, no matter how faithfully these poor devils wore their plastic forms.' There was not enough penile skin and there were problems with blood supply, so he was keen to see the results from Casablanca.[51] Belt and Benjamin were initially positive: they were very impressed with the vagina of one of the patients they referred to Burou in 1962.[52] Benjamin presented another trans woman and her vagina to a lecture at the Albert Einstein Medical College in 1964: no wonder trans people have been hostile towards the medicalization of transgender. 'She was quite a sensation', he reported to Belt. 'Five or six urologists and gynecologists examined her, and were as pleased with the result of her operation as you are.'[53] Yet the duo had reservations a year later when they assessed another example of Burou's surgery, where the vaginal canal was simply too small.[54] Their correspondence indicates a growing scepticism about the penile inversion method. By 1968, Belt was reporting the case of one of their patients injured by Burou and 'left with a recto-perineal fistula and a stenosed [narrowed] urethra with a completely atretic [closed] vagina'.[55]

Belt performed transsexual surgery from 1953 until his family asked him to stop in 1962: twenty-eight operations in all, detailed in a letter of 15 April 1963. His first, outlined earlier, was the removal of the penis and vaginoplasty, but most of his procedures consisted of the removal of the urethra, the transplantation of the testicles into the abdomen, and the plastic construction of a vagina in the perineum. (No female-to-male surgical patients were listed.)[56] He retained the testicles because he thought that their removal would cause endocrinal imbalance, though it was not a method favoured by Johns Hopkins ten years later, because, they informed one of Belt's patients, they thought that it could cause baldness and cancer of the testicles.[57] During his decade of surgery, Belt was dependent on the good will of hospital superintendents and staff and the behaviour of patients. Hollywood Hospital was sympathetic to his procedures in the late 1950s.[58] After 1962, he still consulted transgender patients and conferred with Benjamin, as we have seen, but would send them elsewhere for surgery.[59] By 1966, Benjamin was observing gender reconciliation surgery at Johns Hopkins and implied that he and Belt now had somewhere to refer their patients. The Johns Hopkins surgeons used both penile and scrotal skin for the vagina. Benjamin was clearly impressed by what he termed their 'sympathetic attitude toward the transsexual phenomenon'.[60]

The endocrinologist Benjamin was by no means an unequivocal advocate of transsexual surgery. However, when the need was there, and he was convinced that the candidate was a genuine case, he was determined in obtaining the required bodily modification. In the late

1950s, Benjamin sent one of his 'female transvestites [*sic*] who lives the life of a man' to a private clinic in Beverley Hills where they had their breasts removed: 'I understand from that patient that he operates in a well-equipped operating room in a private home with facilities for patients to stay there like in a hospital.'[61] Indeed, Benjamin set out the benefits of operating in one of his earlier letters to Belt: surgical, psychological, and practical. The first, he said, flattering his correspondent, needed no discussion as the surgical outcome would be the result of Belt's skills. The second, the 'psychological outcome', was 'a chance to acquire a psychological satisfaction and, therefore, a happier life due to the removal of the hated male organs [he was referring to trans women] and due to the realization that they have come as close to the female sex as medicine can provide'. 'These patients', he continued, 'had already experienced a psychological benefit during the period of hormonal feminization. Unless I have been able to observe such benefit, I would not have consented to the operation . . . Since the mind could not be adjusted to the body, the body was adjusted to the mind.' The practical outcome, the third, was 'the prospect of producing a reasonably successful "woman"'. Here there was a hint that not all prospective women were equal: 'In this respect, the physical structure and appearance of the patient is of importance.'[62]

In 1965, Ira Pauly summarized the state of play regarding what he termed 'a dramatic psychiatric syndrome', 'the male's [*sic*] intense desire for sexual transformation by surgical and/or hormonal means, based on his [*sic*] complete identification with the feminine gender role'.[63] Pauly, Professor of Psychiatry at the University of Oregon Medical School, was able to draw on a clinical literature of over fifty European and American studies, mostly from the 1950s and early 1960s, and argued that transsexualism represented medical advances in surgery and hormone therapy making possible 'the partial realization of fantasies of sexual metamorphosis'.[64] He counted a total of 603 reported 'male' and 162 'female' transsexuals, who, he argued, represented but the tip of an iceberg. As early as 1965, then, it was recognized in academic/medical circles that there were both male and female transsexuals.[65]

In 1968, the same author reported on the post-operative adjustment, socially and emotionally, of 121 male-to-female transsexuals, concluding that it was overwhelmingly successful. Hence, he reasoned, and given the relative failure of psychotherapy to revert rather than facilitate the process, 'the apparent success of sex reassignment surgery . . . compels one to accept the surgical treatment of transsexualism on an experimental basis until the initial results can be verified or contradicted or until alternative treatment procedures prove successful'.[66] It was, in other words, cautious psychiatric support for transsexual surgery.

At a Popular Level

We have already noted the shifting attitudes in *Sexology* magazine during the 1960s, with its provisional acceptance of a surgical intervention barely conceived by its editors in the 1950s. Thus, when F. S. from Indiana ('I am what the world terms a transsexualist') sought advice about 'change of sex' in 1964, the outcome was more positive: 'There are . . . a number of cases now in this country where the operation to change a man into a woman has been performed. Some of these have turned out quite successful, with the woman actually marrying as a woman after the operation, and finding that it was possible to carry on reasonably normal sexual intercourse.' (Though it was also pointed out that the procedures were costly, and that few surgeons actually performed the operation and fewer hospitals permitted it.)[67] *Sexology*'s change of heart reflected the reality of medical developments, but it also signified the influence of Benjamin. The same advice column referenced the expert's work and an article that he had written for a previous issue of the magazine.[68]

But, regardless of the magazine's attitudes, the significant factor is its readers' growing quest for information. 'Dear Doctor: Do the men who have themselves castrated and have artificial vaginas constructed learn to reach orgasm in their rôle as women?'[69] 'Dear Doctor: Is there any way possible for a woman to be converted to appear as a man – so much so that she could marry a woman? I am 5' 1½" tall and would like to be at least 5' 8½".'[70] 'Dear Doctor: I am a female of twenty-one years, in desperate need of your help and advice . . . I believe I am a transsexual. Can you advise me as to where I can have sex change surgery performed?'[71] Transsexuality was taking hold at a more popular level.

This is reflected in a series of transsexual pulps too, for Jorgensen was merely among the earlier of several trans celebrities, even if – for some – the fame was short-lived. Hence, Abby Sinclair's *I Was Male!* (1965) was billed as making Jorgensen's story 'read like Dick and Jane'.[72] It related her career in modelling and exotic dancing, gender reconciliation in Casablanca, glamorous life in Paris (a guest of Roger Vadim and Brigitte Bardot), and engagement to a New York lawyer. The book's message was that Sinclair was one of many: 'There is a new species of being on the rise in the United States today . . . Your secretary, your next door neighbor, your girl friend may once have been a man.'[73] Hedy Jo Star's *My Unique Change* (1965) told the story of another stripper and dancer, 'the first sex change to be performed in America':[74] 'Today I am a whole woman. With the help of sympathetic doctors my body has changed until it gradually became less and less a prison and more and more a reflection of the woman within. Then, through surgery, I was physically made a woman.'[75]

Some of these women were the stuff of fantasy. Coccinelle, a French celebrity whose fame became international, and who was sometimes confused with the French actress Brigitte Bardot, was considered one of the most beautiful women in the world: 'that figure with its tantalising walk and diabolically feminine appearance, evoking an atmosphere of sensuality that was almost unbearable'.[76] But the twist in her 'Strictly Adult Sale Only' biography, with its more than sixty plates, was that 'this woman, this superbly beautiful woman, is in actual fact, not a woman at all but – a MAN'.[77]

The interaction between the medical experts and the trans community is intriguing. *Turnabout* referenced work by Magnus Hirschfeld, Havelock Ellis, and David Cauldwell, warning its readers away from 'a growing body of unscientific trash calculated to earn its publishers a fast buck'.[78] Its editor summarized an article in the *British Medical Journal* and discussed the early research of John Money.[79] *Turnabout* ran articles written by Benjamin and other authorities, including Benjamin's 'Advice to a Transsexual' (1963), reprinted from *Sexology* magazine, 'in the belief that it is of prime importance to those among our readers who may be considering a "sex change" operation'.[80] In fact, Benjamin wrote to *Turnabout* urging tolerance when some transvestites mocked transsexuality, and he referenced the magazine in his book *The Transsexual Phenomenon* as providing 'self-expression for their readers through letters and photographs' – which, indeed, it did.[81] Benjamin circulated the abstract of one of his papers a month before it was delivered to the Sixth Annual Conference of the Society for the Scientific Study of Sex in New York in 1963.[82] An observer from *Turnabout*, its indefatigable editor, reported back critically from that very symposium.[83] And a declaration by Lorraine Channing – 'Each day I live a lie . . . Physically I am a man; mentally and emotionally I am a woman. I am a transsexual' – ended up in the pages of *The Transsexual Phenomenon*.[84] There was interaction with the wider trans community that extended beyond the consulting room.

Benjamin was also involved in support groups. We have the somewhat gossipy notes of transsexual support group meetings in New York in the late 1960s: 'Next meeting will be . . . in Brooklyn. If anyone says it's a depressed area, you're right. It's a depressed area, but no one is depressed there.'[85] The trans women group compared operational procedures and surgeons: Burou in Casablanca, considered the most experienced; a Cuban surgeon who operated in New Jersey and Rome and who attended one of the meetings; the previously mentioned Belt, 'the biggest butcher of all. He's destroyed more people.'[86] There was post-operative comment, too. Janet said that men mistook vaginal tightness for virginity or nerves.[87] Phyllis, who attended with her husband, and who claimed to have undergone the first transsexual operation at Johns Hopkins, complained about the dimensions of her new vagina.[88] Though these notes are in Benjamin's archive, they are a reminder of the independence of patients – as in the scathing verdict on Belt – and we will see later that *Turnabout* was critical of aversion therapy.

Benjamin's papers also contain examples of patient agency. One of the subjects of the earlier-referenced 1955 report in the *Journal of the American Medical Association* wrote an angry letter to the journal's editor, complaining that the authors had not sufficiently represented the homosexuality of their cases, had asked directed questions, and, as she put it, had 'twisted' her words 'to suit their purpose'. 'Had the authors made a genuine attempt to establish a rapport with their subjects, instead of attempting to milk scientific information from them in the approximate manner laboratory animals are used, results would have proven far different.'[89]

Treatment

There is no doubting the rapid expansion in trans-focused facilities. Dave King has referred to the world of the transvestite being super-seded by the world of the transsexual in a 'mere five years' in the second half of the 1960s.[90] By the time they presented to the annual meeting of the American Psychiatric Association in 1977, Jon K. Meyer and Donna J. Reter were able simultaneously to assume the 'normalization' of sex reassignment and cast considerable doubt on the outcomes of its 'almost routine acceptance', an important matter that we will return to in due course.[91]

As we saw in the previous chapter, the history of transsexual surgery is closely linked with treatment of the intersexed – the chronologies are identical – where, from the 1950s, a protocol was established for clinicians to determine the sex of infants born with a 'gender-atypical anatomy' (of doubtful sex) so that their bodies could be modified to correspond to the assigned sex.[92] Though the intersex patients (infants and children) were then younger than transsexuals (always adult when cutting was involved), surgical procedures overlapped. Genital surgery for the transsexed did not involve clitoral reduction as it did for the intersexed, but penectomies (removal of the penis) and vaginoplasty (construction or modification of the vagina) were performed on both types of patient. Robert Stoller, Professor of Psychiatry at the School of Medicine, University of California at Los Angeles, and a well-known transgender expert by the 1970s, was reporting on intersex surgery in the 1960s, outlining cases of created vaginas and (badly) constructed penises.[93] Johns Hopkins was a pivotal institutional centre in the diag-nosis and treatment of both intersex and transsex, and the role of John Money features in both histories.[94]

After the initial reluctance of the late 1940s and 1950s, the list of possibilities for transgender treatment during this period expanded rapidly. The Erickson Educational Foundation, established by Reed Erickson, an eccentric, millionaire trans man, provided a source of research funding and public information on transgender matters throughout the 1960s and 1970s, including contact with sympathetic medical professionals.[95] We know that teams of surgeons and psy-chiatrists were offering transgender evaluation, treatment, and sex reassignment surgery in the 1960s: at the Gender Identity Clinic of The Johns Hopkins Hospital; the Department of Surgery at the Stanford University School of Medicine; New York's Payne Whitney Psychiatric Clinic; the Gender Identity Research and Treatment Clinic at UCLA; the University of Michigan; and the Passavant Memorial Hospital of Northwestern University Medical School in Chicago.[96] The University of Minnesota instituted a Transsexual Research project in 1967, providing free surgery for male-to-female patients, mostly from the Twin Cities of Minneapolis–Saint Paul, in return for the subjects' participation in their enquiry.[97] The trend continued into the 1970s. Benjamin's archives contain 1970 lists of both Gender Identity Clinics (fifteen in number) and institutions and surgeons 'who have or are performing sex reassignment operations' (ten in number), including several private, sole practitioners in New York and San Francisco.[98]

It is difficult to provide a realistic count of the number of patients in these programmes, given the unsystematic nature of reporting and the constantly shifting timelines. We know that Johns Hopkins saw 87 patients from 1969 to 1972, only 22 of whom were genitally reassigned.[99] In a paper published in 1982, Meyer said that he and his colleagues at Johns Hopkins had seen over 500 patients in ten years 'who wished to have their genitalia and other physical attributes modified so as to remove a dissonance between sense of self and physical body'.[100] The Gender Dysphoria Clinic at the Case Western Reserve University School of Medicine in Cleveland, which favoured psychotherapy rather than surgery, treated over 200 transgender patients from 1974 to 1982.[101] The Gender Identity Association in Jacksonville, Florida, a non-university, private-practice team, though modelled on university lines, had by 1973 evaluated 17 male and 15 female patients and operated on 12 – 4 male-to-female and (interestingly) 8 female-to-male transsexuals.[102]

There was a noticeable gap between the visibility of transsexuality and the scale of demand for such consultations, on the one hand, and the reality of the relatively small number of actual gender reassignments, on the other. Meyerowitz was probably realistic in her calculation that, by the mid-1970s, about twenty major medical centres were offering treatment, but may have exaggerated somewhat in her estimate that some thousand transsexuals had been provided with surgery.[103] Stoller, who was in the know, said in private communication in 1970 that demand outstripped supply. The

> grim reality in regard to transsexualism . . . is that there are thousands of patients requesting evaluation and treatment and only a handful of physicians familiar enough with the condition to be able to do so. When you add to this the problem that almost no sex-transformation operations are being done out in the open in the United States (though there are an unknown number being bootlegged to those who can afford them), the problem becomes very severe.[104]

Only 70 (22 per cent) of over 300 patients in Benjamin's early files (up to 1966) had had a 'conversion operation', though many more had 'contemplated' surgery and most were being administered hormones (hence their presence in Benjamin's dossiers).[105] We can dismiss the journalistic claim in 1980 that '10,000 Americans . . . have achieved transmogrification'.[106]

Then there is what Deborah Heller Feinbloom called 'a kind of underground similar to the abortion situation some years ago' (she was writing in 1976). Surgeons in smaller hospitals might perform surgery with minimal psychiatric evaluation and few questions asked, provided the money was paid.[107] Writing in the same year, Stoller told a correspondent that hormonal and surgical treatment was not difficult to get if a person had money: 'certain private practitioners are clearly more interested in collecting large fees than in scrupulous practice. I think it is safe to say that in the States, anyone with enough money can get the operation, no matter what his psychological condition, those without money are practically guaranteed not to be able to receive treatment.'[108] Benjamin's 1970 list of those providing gender reassignment surgery had comments such as 'Reported to charge $5,000 and will operate on almost anyone' and 'Very hush-hush'.[109]

As was discussed in a previous chapter, self-help must have been ubiquitous, given the inequality mentioned by Stoller. In 1963, *Turnabout* magazine found it necessary to warn its transsexual and cross-dressing readers away from purchasing their female hormones from the animal nutrition section of the Sears Roebuck Catalogue.[110] A correspondent in *Sexology* in 1965 referred to 'buying female hormone facial cream and eating it as well as lubricating my whole body with it'.[111] The 1980 journalistic estimate claimed that less than half of the estimated operations were 'documented' and the remainder were in 'back street clinics or abroad'.[112]

The early focus was on male-to-female transsexuality, those whom we now term trans women. But research, discussion, and surgery increasingly included the female-to-male: the trans man. Benjamin's *Transsexual Phenomenon* had just one chapter on female-to-male transsexuals, based on 20 treated cases, compared to the 152 male-to-female transsexuals who were the primary focus of his study.[113] Psychiatrists from the Mount Sinai School of Medicine of the City University of New York published a case of 'female transsexualism' from the 1960s. The patient had first been diagnosed as a depressive homosexual but had denied any homosexuality and pressed for surgery. 'In early 1966', Gloria Warner and Marion Lahn reported, 'the patient appeared as a small, ruddy, slightly effeminate male, who worked successfully as a bookkeeper, lived as a husband and father.' It seems from the report, published in 1970, that Mount Sinai's role was limited to psychiatric assessment and that surgery was (persistently) sought elsewhere: 'Clearance was necessary before the gynecologist she [*sic*] had consulted would proceed with bilateral mastectomy, complete removal of internal genitalia, closure of the vagina, and hormonal treatment for clitoral enlargement.' The trans man, from South America, had initially travelled to the USA specifically for surgery.[114] A 1968 report from the Gender Identity Clinic of the Johns Hopkins Hospital noted that, of the hundreds of requests for evaluation for sexual reassignment surgery received by the clinic (where assigned sex at birth was known), about 80 per cent were from 'males' and 20 percent from 'females'.[115]

Clearly, then, trans men were presenting in the 1960s and 1970s. The New York psychiatrist Ethel Spector Person was treating trans men in the early 1970s. One said that he was annoyed by physicians who refused to countenance the idea of a male without a penis: 'It's comical, it's absurd to think that if you have no penis you're not a boy ... I felt they were taking away my masculinity.'[116] Pauly published a two-part article on 'female transsexualism' in 1974, drawing on an already substantial international literature of eighty such cases and his own clinical experience involving the fifteen Oregon trans men whom

he had seen in the previous ten years.[117] Despite his earlier reports of
the heavy transsexual gender imbalance, he felt by 1974 that the distri-
bution was more equal: about 2 or 3 to 1 of MTF/FTM, but possibly
approaching a 1:1 ratio.[118]

Pauly compiled what he termed the 'natural history' of the trans
man.[119] They had, from their earliest recollections, a preference for
the masculine: they played the role of tomboy; they cross-dressed;
and, after unwelcome puberty, they frequently bound their breasts
to make them invisible. They were sexually attracted to females, but
though they might experiment with lesbianism, usually viewed attraction to females as heterosexual rather than homosexual, given their
core masculinity. The partners of these transsexuals related to them as
men, and it was not uncommon for them to be unaware of the designated sex of their sexual partner until their transsexuality was declared:
'these young girlfriends of female transsexuals had never been previously involved in homosexual relations, had experienced heterosexual
social and sexual relationships with pleasure, and preferred to respond
to their transsexual lovers in a heterogenderal, if not heterosexual,
manner'.[120]

A Seattle study in the early 1980s, supported by a grant from the
Erickson Educational Foundation, reported that almost 30 per cent of
the sixty transsexuals who visited the clinic in an eight-year period were
'women' seeking male sex reassignment.[121] The treatment outlined was
conservative: detailed evaluative interviews; monthly psychotherapy for
a minimum of a year; and then, if approved, hormonal treatment. The
final stage, for the undeterred, was surgery, most usually mastectomies. It is clear from this study that not all those who first presented
completed full transition. Of nineteen original 'female' patients, fifteen
made it through the initial evaluation, but only nine had embarked on
hormone therapy and just five of those had had mastectomies. Only one
female-to-male transsexual proceeded to phalloplasty (the construction
of a penis). Nine of the original 'women' were living as males at the
time of follow-up research; four were lesbians; and two were 'in a less
clear-cut male/female role'. One of the 'women' with a mastectomy and
hormone therapy lived as a male for two years but then reverted back
to her 'female' identity. The authors of the study stressed the variety
in these sexual histories – the 'persistence of a certain brittleness of

gender identity', as they termed it – and hence the need to proceed with caution in sex reassignment matters.[122]

A comparative stereotype emerged. Female-to-male transsexuals were characterized by their 'undramatic profile', as another Johns Hopkins study put it.[123] They quietly passed, in contrast to the flamboyant 'posturing' of male-to-females.[124] Their sexual repertoire was rather limited compared to a heterosexual control group. Indeed, the researchers were taken with their subject group's relative normality: 'a majority of them have had intimate sexual relationships with males at some time in their lives'.[125] But there was no consensus. One of the 'myths' challenged by Leslie Martin Lothstein's study of female-to-male transsexualism was that 'female transsexuals [*sic*] are more stable than male transsexuals [*sic*]'.[126]

Lou Sullivan

Much of the material on the transsexual moment is based on the medical literature, though I have consulted trans publications and a popular magazine. However, we also have the diaries of someone who lived through that period and who interacted with the various custodians of trans realization. Louis Graydon Sullivan, who was born Sheila Jean Sullivan in Milwaukee (Wisconsin) in 1951, was trying to make sense of his desires and identity in the late 1960s and 1970s.[127] In 1973, Sullivan went to a lecture by Christine Jorgensen, but recorded that it was difficult to go because it was a public declaration of being interested in 'this male/female thing'.[128] He researched cross-dressing, observing that Esther Newton's study of drag queens, *Mother Camp* (1972), had 'a beautiful picture of exactly what I want to be: it was of Desirée, a female impersonator who honestly looks exactly like a beautiful young woman, figure & all. In the picture he was naked except for a G–string & he looked like a beautiful boy-woman.'[129] Sullivan experimented with transgender: a panty girdle over the breasts to flatten them and a stuffed sock or glove in the crotch. Sullivan said that drag was liberating because it made a person feel that masculinity and femininity were not natural, 'that it's *not* in anyone's *nature* to be masculine or feminine'.[130]

The future gay trans man kept trying to make sense of his feelings. 'I don't feel like a "male trapped in a female's body" [one of the conventions of transsexuality] nor do I think I *could* be a man. It's the fantasy . . . if I had a mastectomy I'd have to beat off 24 hours a day because I'd be so turned on by myself. The whole scene for me is just a wonderful sexual fantasy.'[131] Yet he is contemplating more radical bodily reassignment and has clippings regarding 'sex change' surgery (another term used then and relayed in the diary):'I'm really becoming uncomfortable in and tired of my ambiguity.' The 'fantasy' was giving way to 'intense desire' to act.[132]

Sullivan's archive contains a copy of his (unsuccessful) application to the Stanford University Gender Dysphoria Program in 1976. The answer to their question regarding 'her' preferred method of sexual contact was likely to have raised some concerns in what was, as we will see, an essentially cautious experiment: 'I want to be a gay male having sex with another gay male.' It would have done nothing to reassure therapists who, in respect to what they described as female-to-male transsexuals, were looking for masculine (heterosexual) attractions to women rather than (homosexual) longings for men. Lou's was hardly the sort of profile that the Stanford team would have been seeking.[133] It was in that very year that an author of a study of transsexuality stated that she was 'unaware of any female to male transsexuals who consider the possibility of male homosexuality'.[134]

It was no inexorable trans journey for Sullivan, then, who, to return to 1960s and 1970s terminology, oscillated between transvestism and transsexuality. He became concerned by his inability to fit in anywhere. As he recorded, why would the gays accept him as a woman? Feminists did not accept his dress. Lesbians? 'I like men.' The straights? 'No way.' Transvestites? They were all male-to-female rather than female-to-male. He told his counsellor that he was past the 'crisis of wanting a sex change'.[135] In 1977, he wrote to a friend that he was 'growing more & more comfortable with myself as Sheila – able to relax more as a female who digs sometimes passing/looking like a male. I'm giving myself permission to be a female transvestite.'[136]

But the homosexual longings continued. Lou remained unclear about his identity. 'I don't believe I *am* a man, but I am certain I *want* to be one', he wrote to Mario Martino, the trans man autobiographer and gender counsellor, in September 1979: 'I don't know if I am a deluded transvestite or an overly-cautious transsexual.'[137] Towards the end of

1979, and after saying that he was not transsexual, Lou wrote, 'I'm taking the big step, I'm going to begin taking male hormones & live fulltime as a guy . . . It came because I no longer want to live in a fantasy world . . . I'm tired of pretending.' His counsellor was encouraging:

> I told him I didn't feel like a 'man trapped in a woman's body' & he said nobody did, that was just some catchy phrase the medical profession made up. I told him that I didn't fit any of the stereotypes of a TS & he said that being a TS does not dictate anything about you other than the way you feel about yourself & that I have a perfect right to be a gay man if that's what I want.[138]

Transvestism

It is vital to emphasize that what has been termed the transsexual moment consisted of a lived culture. It was not mere medical history. Moreover, the transsexual moment could equally have been termed the transvestite moment, given the visibility of cross-dressing throughout this period. In 1966, Darrell Raynor thought that transvestism was a 'growing American phenomenon, that it is more widespread than any imagine, and that it will become more so'.[139] As men took refuge from the 'atomic abyss' in the momentary diversion of femininity – it is not clear what sanctuary their wives and girlfriends had – it almost seemed that America would turn trans.[140] After a particularly well-attended gathering at the cross-dressing venue Casa Susanna, and in the company of the academics Hugo Beigel and Wardell Pomeroy, Raynor mused that there were enough ex-navy men there to 'crew and command a small war vessel', sufficient army and marine veterans to 'take an enemy machine-gun nest'.[141] Many transvestites were engineers, he wrote, and he speculated, stretching logic somewhat, how much of America's space programme had been due to 'transvestite engineers and technicians'.[142]

With their photographic vignettes, drawings, advice columns, editorials, fiction, and cross-dressing and beauty tips, *Turnabout* and *Transvestia* magazines provide a glimpse of this culture of cross-dressers. *Transvestia* provided long tips on cosmetics, foundations (heavier for the hirsute man), powder, rouge, eye make-up, lipstick, nail polish (avoid 'wild shades'), creams and lotions, perfume ('perfume always enhances a lady'), and deodorant. Readers were advised on how to minimize their masculinity: avoid drawing attention to large hands, feet, and arms, either through the use of neutral colours so as not to draw attention or by the use of loose clothing; disguise the Adam's apple with dark foundation and head posture ('keep the chin down') and do not draw attention with jewellery; avoid an obviously artificial falsetto voice (women have 'voices as deep as many men'); shave with an electric razor and then a safety razor; wear flesh-coloured tights or even two pairs of nylons if the leg hair becomes unmanageable; do not pluck one's eyebrows pencil thin (no longer fashionable with women). And there were further instructions for lady-like behaviour. Never stand or sit with legs apart. Always go through the door first in the company of men. Avoid being too intelligent in conversation, not because women lack intelligence but because they know how to make men feel superior. 'Learn to be socially passive and acquiescent – it is difficult, but necessary.'[143] There were jocular, in-house stories about unsuccessful gender presentations: the overweight, hairy neophyte in a tight dress, with flesh and 'black fur' spilling out everywhere. 'I cannot conceive a TV without a certain esthetic sense', wrote Susanna Valenti in her 'Susanna says' column.[144]

There was, of course, a direct relationship between such magazines and the lives described between their covers, between text and community. As Robert S. Hill has explained, 'These crossdressers assembled first in a storytelling, textual space and later in organized social groups.'[145] The photographs in these magazines were part of much wider – affirmative – collections of men in female dress, of the kind found accidentally and published in the book *Casa Susanna*.[146] See Illustration 8. The sharing of photographs was an important part of group socialization.[147] 'These people aren't only collecting photos', writes Ms Bob Davis, 'they're networking, making personal contacts, and constructing a femme identity'.[148] There was, then, a shift from sometimes isolated, often secretive, gender expression to the formation of a more outwardly expressed, collective sense of self: from a 'private practice' to a 'social identity', as Hill puts it.[149]

Sophie Hackett has explained that the photographs were important in themselves, providing 'a chance to bring that woman to life in the particular way that only photographs can'.[150] In one, the women

8   *Attributed to Andrea Susan, 'Susanna at Casa Susanna', 1964–9.*

all hold cameras, pointing at one of their short-skirted number in a paparazzi-like pose. The message is that these women are to be looked at. See Illustration 9. The photographs were endorsements of a range of femininities, respectable mainly (more of which soon) but with a hint of stocking top, here, a lifted dress, there, a negligée, a feminine pose. They were available for private confidence building (instant gratification in the age of the Polaroid), or public acclaim if published in the transvestite magazines.[151] It is probably difficult for the non-cross-dressing, modern viewer to understand this imagery totally, given the double distance of identity and time, but they still carry sufficient appeal to be included in a 2018 photographic exhibition in London's Barbican, and in the accompanying art book, *Another Kind of Life: Photography on the Margins*, appearing after Diane Arbus and Bruce Davidson, and before Danny Lyon and Mary Ellen Mark.[152]

The dominant femininity of these cross-dressing, heterosexual, usually married, middle-class men was a respectable womanliness – with, we have noted, a hint of mischief.[153] It was this aspect of the

9   *Attributed to Andrea Susan, 'Photo Shoot', 1964–9.*

photographs that attracted *Casa Susanna*'s editor Robert Swope when he found them discarded in a fleamarket: 'typical, middle-class suburban women . . . housewives who know how to dress up for a night out, and certainly don't mind having a drink'.[154] 'The general tendency among *Transvestia*'s crossdressers in the early 1960s', Hill writes, 'was to de-sexualize transvestism'.[155] It was a public persona, cultivated to lend respectability to the cause, and comparable to the reputable face of organized homosexuality presented by the Mattachine Society.[156] Privately, the asexuality was not as marked, as indeed the photographs suggest. Virginia Prince, the editor of *Transvestia*, seems to have been unusually candid about her desires in her extended interviews with the gender researcher Stoller.[157] This said, when Pomeroy and Beigel visited Casa Susanna, they were surprised at the lack of 'allusions to sexual matters': 'These people could have been recommended to children as examples of middle-class decorum.'[158]

There are amusing discussions of the distinction between erections as a result of female appreciation and mere fetishistic attraction to an object with female associations.[159] The twinsets and pearls and bouffant hairstyles of *Transvestia* and *Turnabout* (and presumably their readers too) were poles apart from the paraded, exaggerated femininity of the drag queens, from whom they were so keen to distinguish themselves – the very contrast that appealed to Swope.[160] Raynor's *A Year Among the Girls* quoted Virginia Prince as saying that the 'homosexual queens' gave transvestites a 'bad name': 'These queens conduct themselves as harlots, but transvestites identify with ladies and carry themselves with propriety.'[161] *Turnabout*'s review of John Rechy's iconic, drag-queen novel *City of Night* (1963) turned its metaphorical nose up at 'a dangerous book': 'we hate to think of the misunderstanding it may generate . . . particularly if transvestites can be lumped together with male homosexual prostitutes'.[162] Jean Genet's *Our Lady of the Flowers* (1963) was likewise dismissed as transvestism of the wrong kind.[163]

The literature that these heterosexual cross-dressers produced and consumed – their fantasy world – had a naivety about it. Those who have read scores of these novels and short stories produced during the 1960s and 1970s have noted its purity: the 'stereotypical story . . . pictured the transvestite as a very nice person – actually more of a Barbie doll than a person'.[164] The cross-dressers of in-house fiction are innocents, forced or persuaded into their transvestism.[165] Stoller said that he found it difficult to read the material because of its boring repetitiveness, but that transvestites were highly aroused by it – it was their pornography.[166]

The heterosexuality of these male-to-female cross-dressers was emphasized, especially in Prince's *Transvestia*. Its goal, outlined in its opening pages, was to cater for the 'needs of those heterosexual persons who have become aware of their "other side" [femininity] and seek to express it'. While it did not 'condemn nor judge the fields of homosexuality, bondage, domination or fetishism . . . [t]hese are left to others to develop. They are not part of the areas of interest of this magazine.'[167] *Transvestia*'s prototype, a typed newsletter from 1952, had been blunter still: 'there is necessarily no identification with or similarity to the sex deviate commonly known as homosexual. As a matter of fact, there is even a wider disparity between homosexuals and transvestites than between homosexuals and the "normal" heterosexual individual. Transvestism should not be confused with or compared with sex deviates.'[168] Accounts of the heterosexual cross-dressers emphasized the masculinity of the men under the dresses: 'His voice was deep and masculine and his handshake was strong. No faggot this'; 'This was a strong, healthy young man, about thirty years of age, well-muscled, and very much a he-man.'[169] As Stoller once expressed it, the transvestite ultimately 'believes himself a better woman than any woman because he is the only woman who surely has a penis'.[170]

The public visibility of transvestism was evident in books like Edward Podolsky and Carlson Wade's *Transvestism Today* (1960), reviewed amusingly by *Turnabout* as containing 'the most banal and imbecilic case histories ever fashioned'.[171] Raynor's *A Year Among the Girls* is a product of this period too, but there were many other such publications. *Sex Life of a Transvestite* (1964) invoked the authorities (Benjamin, Prince), amidst pulpish, first-person confession: 'Her naked breasts rose and fell invitingly as she took each ragged breath. I wondered what it must feel like to have breasts like that, so large and yet so lovely.'[172] Books like that argued for the normality of cross-dressing while demonstrating its opposite.[173]

If the public presentation of transvestism in *Turnabout* and *Transvestia* was as a kind of asexuality, the pulp literature smuggled sex back in the form of autoeroticism. 'Not all masturbators are transvestites', proclaimed *Transvestism* (1969), 'but it would be safe to say that nearly all transvestites are masturbators'.[174] And it quickly followed with a case of a boy – 'History of M.: The Penis Doll' – who masturbated using his sister's doll's dress or a glove puppet to clothe his penis![175] A cross-dressed penis?

There were numerous cheap pamphlets that seemed to cater for more specialized tastes. At $3.50 each, close to $30 in current terms, Nutrix Corporation and Candor Books offered a range of such fare in the 1960s. Their series, *Letters from Female Impersonators*, purportedly genuine letters describing episodes of cross-dressing, was into volume 15 by 1964. The descriptions in that volume were accompanied by photographs of some very masculine-looking women, including the attractive, though seemingly hirsute blonde on its front cover (blonde wig and black chest hair), and the sinewy Betty Bill, who looks remarkably like the Rolling Stones Bill Wyman, when he was much younger.[176] Nutrix (as in new tricks) offered the multi-volume *Art of Female Impersonation* as well as bondage and spanking pamphlets; their repertoire can be followed in the Library of Congress Catalogues of Copyright Entries for the 1960s, as well as in the recorded exploits of Nutrix's owner Irving Klaw (of Betty Page fame), who was prosecuted for sending obscene material through the mail.[177] Only a minor part of Klaw's empire, Nutrix's female impersonator offerings appear a far cry from the busty allure of Page, a famous 1950s female bondage model, although both came together momentarily in two of Klaw's films, *Varietease* (1954) and *Teaserama* (1955), where the female impersonator Vicki Lynn appeared with Page in what Eric Schaeffer has termed a classic case of 'gender sabotage'.[178] That the tease in *Varietease* and *Teaserama* consisted of a heterosexual pin-up model and a cross-dressed man – with appeal to an ostensibly straight audience – indicates (notes Schaefer) that an unexpectedly polymorphous desire could lurk at the heart of such burlesque.[179] Page and the female impersonators were not so different from each other after all.

Female Impersonation

We have so far circled around the drag queens – or the female impersonators, as they were then known and as whom they self-identified. They have only been encountered as an 'other' to the heterosexual cross-dresser. But they too were part of the moment discussed in this chapter. They belong more in the homosexual milieu than in the more conventionally heterosexual community of the *Transvestia* cross-dressers – we have already noted the hostility of the latter to the former. Minette's *Recollections of a Part-Time Lady*, with memories of performing in drag from the 1940s through to the 1960s, is pitched in terms of knowing references to seafood (sailors), gay people, campy people, browning, oral sex, and fairies ('I was a fairy').[180] Seattle's 1940s and 1950s cabaret club Garden of Allah was a location that offered a home to a range of homosexual and heterosexual sexual and gender identities: male and female sex workers and lesbians as well as masculine trade, women in male dress as well as men in women's clothing, feminine-identified males, and future trans women: 'The messages of the strip were that there was humor in sex, that being sexual was fun and natural, that it pushed the limits of community tolerance, and that the impersonators had the courage to push those limits.'[181] A few of the performers in the Seattle cabaret in the 1940s and 1950s saw themselves as transsexual in the 1960s.[182] But many seem to have identified as gay men whose femininity gave them sexual access to the straight men that they desired. There are the remembered stories of lumberjacks and of sailors' neckerchiefs hanging from chandeliers.[183] The Garden of Allah had a reputation for its blow jobs, even if the reputation was misplaced.[184] It is no accident that Don Paulson and Roger Simpson's oral history of the club appeared in a lesbian and gay studies series.

Female impersonation was pervasive in the 1950s and 1960s at the very time that Benjamin was formulating his notions of transsexualism. In fact, Laura Grantmyre has shown that in Pittsburgh's Hill District the staged cross-dressing of the 1920s and 1930s continued into the later period, captured in the compelling photographs of black female impersonators now in the Carnegie Museum of Art.[185] See Illustration 10. Such acts featured in nightclubs and bars throughout the 1950s and 1960s in this part of Pittsburgh, and Grantmyre's oral history demonstrates that these gender-crossing performers formed part of the local African-American community.

Thaddeus Russell has noted the persistence of an earlier African-American toleration of gender flexibility, of drag balls and female impersonation acts, in many American cities into the 1950s, and of the willingness of black newspapers and magazines to cover such activity.[186] *Ebony* reported on masquerade balls in Harlem (Fun Makers) and Chicago (Finnie's Club) in 1952, attended by hundreds of 'mimics'. 'Female Impersonators Hold Costume Balls' featured a photo spread of 'well-turned' legs and lavish coiffures, poses with the obligatory sailor, and shots of the cross-dressed at the entrances to both the men's and women's restrooms.[187] 'So expert are some of the impersonators at mimicking not only the appearance but the gestures and voices of women that many of the spectators found themselves unable to believe that some of the "women" were really men.'[188] These were multiracial and inter-class affairs – true intersectionality at work – but with a strong black presence: *Ebony* noted the many 'Josephine Bakers' at the 1951 balls.[189] The point was, as Allen Drexel expressed it some years ago, the 'exceptional openness' of that culture.[190]

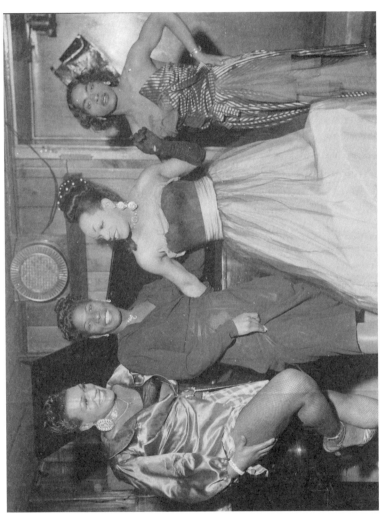

10 *Group portrait of four cross-dressers, including Michael 'Bronze Adonis' Phelan on left, and possibly 'Beulah' on right, posed in front of piano in Granville Hotel, c.1952.*

But other cities were involved too. Sir Lady Java performed in the clubs in Los Angeles in the 1960s. She is pictured in the African-American magazine *Jet* in 1966, 'a go-going bottomless' at the Club Night Life, and then again in 1967 when she protested against LAPD harassment.[191] Female impersonators worked the black clubs in Washington, DC in the 1950s. One interviewed by Genny (then Brett) Beemyn said that, in some venues, they had to wait table and help out in the kitchen as well as the usual singing and dancing.[192] Russell writes of Detroit's 'drag craze' in the early 1950s, in what was then the country's third-largest African-American urban centre: 'Detroit's drag queens were not shy about their sexual orientation. They sang songs of lust for men and flirted with men patrons.'[193] In New Orleans, too, in the 1950s, the Dew Drop Inn hosted female impersonators, captured photographically in a collection at Tulane University.[194] See Illustration 11. The audience and the performers in the pictures are mainly black. Those who strutted their stuff as female impersonators in the famous blues club included Bobby Marchan, who went on to a music recording career, and Patsy Vidalia (born Irving Ale), who was less successful.[195] Both are pictured as vivacious black women on stage at the Dew Drop. Richard Penniman also performed in the blues club in drag as Princess Lavonne before he became Little Richard. As Marybeth Hamilton puts it, he 'dispensed with the drag queen's skirts, but he retained her sequins, her makeup, her pompadour, her strutting self-confidence and her way with words'.[196] In contrast, New Orleans's Club My-O-My was a bastion of whiteness, if the billboards, publicity photographs, and memorabilia are a visual indication. In the photograph of Harvey Lee on stage at the Club, both Lee and his backing band are white.[197]

In Kansas City in the 1950s and 1960s, the Colony Club, Forest Ballroom, Jewel Box Lounge, and many others presented a steady stream of cross-dressing among patrons as well as on stage – and male as well as female impersonation, though the latter outnumbered the former.[198] Esther Newton's major informant for *Mother Camp*, Skip Arnold, was pictured at the Jewel Box in 1959.[199] See Illustration 12. An extensive photographic collection in the Gay and Lesbian Archive of Mid-America at the University of Missouri–Kansas City captures that particular milieu.[200] Like Club My-O-My, and in contrast to several of

11 *Female impersonators, Dew Drop Inn, New Orleans, 1954.*

12   *Skip Arnold at the Jewel Box, 1959.*

the clubs discussed above, the performers and patrons in Kansas City were mainly white.[201]

In fact, female impersonation criss-crossed the nation.[202] The inter-racial Jewel Box Revue, with its twenty-five female impersonators and one male impersonator, Stormé DeLarverié, toured into the early 1970s.[203] Bud Coleman has listed its itinerary, which included both large and small cities: New York, Washington, DC, Cleveland, Detroit, Chicago, Fort Worth, Los Angeles, Rochester (New York), Scranton (Pennsylvania), Akron (Ohio), Cairo (Illinois), Pueblo (Colorado), Boise (Idaho), and Hobbs (New Mexico).[204] The cover to Rae Bourbon's record album *Around the World in 80 Ways* (1956), with its picture of a desultory ageing drag queen, cigarette in mouth, flanked by the legs of two bare-chested sailors, listed an exhausting personal circuit: from the Silver Slipper in Albuquerque, New Mexico, Emmy Wilson's 'Glory Hole' in Central City, Colorado, and Clara's Place in Deadwood City, South Dakota, to the Cheerio Club, Idaho Falls, Idaho, the Coon Chicken Inn in Salt Lake City, Utah, and the Cotillion Room, Tucson, Arizona.[205] Minette recalled that the small-town clubs were 'drag queen crazy' at that time: 'They loved fairies, and they could book us cheap, and we brought business.'[206] As Esther Newton put it, drag was 'as American as apple pie'.[207]

When Avery Willard published his *Female Impersonation* in 1971, then, with its roll-call of stars (Sonne Teal, Lynne Carter, Minette, Chris Moore, G. G. Allen, Robin Rogers, and others), venues (the 82 Club, Jewel Box Revue, Kansas City's Jewel Box Club), and before-and-after photographs and shots of stunning women, he was capturing an important phase in American cultural history.[208]

Drag was not unidirectional. It had multiple readings and appeal. Bourbon's albums and songs, based on his female impersonator acts, relied on double entendre that the audience might miss if they were not in the sexual know – 'Around the World', meaning a tongue bath, for instance. 'A Trick Ain't Always A Treat', recorded at Kansas City's Jewel Box, relied on discerning what a 'trick' was; and 'I Must Have a Greek' (on the album *You're Stepping On My Eyelashes*) referred to anal sex. One can only imagine the stage performance of 'Mr. Wong', the gestures which accompanied the singing, and Bourbon's trademark giggle: 'All the ladies say, Mr. Wong has the biggest tong in the Orient

today. Mr. Wong has the biggest tong in China.'[209] The female impersonator José Sarria, who held court at San Francisco's Black Cat Café
until it closed in the early 1960s, recalled his banter with the audience:
'Young man! Do you have a license to carry that hidden weapon?';
'Now pay attention you handsome man. This is very sophisticated stuff.
You should learn this. You won't always be beautiful, you know. You'll
need something to talk about when the plumbing gets rusted.'[210]

Jackie Phillips, who featured at the Colonial Club in San Jose and San
Francisco's famous Finocchio's, and who knew the circuit of acts whose
venues included the Garden of Allah, explained that some of his fellow
performers revelled in their pretence and that this was the basis of their
audience appeal: 'they loved me . . . because I gave the impression that
"I'm not a beautiful woman. I'm not trying to be a cunt. I'm just a drag
queen and enjoying myself."' Russell Reed relied on comedy. He would
strip and end up in red underwear, shaking 'her' two different-sized
breasts – 'He knew he was fat and he made no bones about it.' Billy
DeVoe, who featured at the Garden of Allah, depended on incongruity:
looking beautiful, like Jane Russell or Jane Powell according to Phillips,
but singing and speaking with a hillbilly voice. Jerry Lane, however,
was a facsimile; she resembled Susan Hayward and 'you thought you
were talking to a girl when you talked to Jerry Lane'. Ray Saunders
was similar: 'She was pretty. And she was feminine! And the guys saw
nothing but cunt in her. Not no boy! No man! Nothing! They saw this
little cunt singing, carrying on. And this was her gimmick.'[211]

The sexualized atmosphere, on and off stage, depended both on
the concealed – though known – masculinity and on the illusionary
femininity. Club My-O-My billed 'The World's Most Beautiful Boys
in Women's Attire', clearly catering for varied sexual tastes. 'The most
interesting women are not women at all', its advertising poster proclaimed.[212] 'Sir', one of Newton's observed performers, retorted to an
unruly, straight member of the audience, 'I'm more "sir" than you'll
ever be, and twice the broad you'll ever pick up.'[213] On stage, Phillips
quipped at his Finocchio's audiences: 'Please don't get excited over
me. I have the same thing you have', or 'I'm sorry darling, I'd tell you
to close your mouth but I don't want to interfere with your sex life!'[214]
Off stage, performers dated men and women as men or women, and
sometimes engaged in sex in the club itself. Phillips recalled those who

masturbated customers under the table when they were being bought drinks. He said that it was treated like a game, 'like hopscotch', keeping score with matches or by shifting bracelets from one wrist to the other (as he did).[215] He had an affair with a married man whose wife did not treat him as a threat in the same way that a relationship with 'another girl' would have been.[216] Sarria quipped of one attractive audience member, 'He could be dinner tonight; I better not insult him.'[217]

We saw in the previous chapter that, in the 1930s, female impersonation on stage merged with cross-dressed sex work, and the pattern persisted into later decades. Minette's memoirs make it clear that sex work was a major way of making a living for her and her fellow performers.[218] Jimmy Callaway, the star at New Orleans's Club My-O-My in the 1950s and 1960s, recalled – in an interview which made it clear that he was gay rather than transsexual – that the performers turned tricks: 'we would get all kinds of propositions from guys who came to see the show'.[219] Vickie Lynn, who worked at the 82 Club in New York in the 1950s, said that, to boost her salary and tips, she 'did a little "modelling" on the side. You know what that means. So that made up the balance in your rent and everything.'[220] It is evident in Newton's classic study of drag queens in America in the 1960s that the female impersonators of the stage liked to distinguish themselves from the street fairies who were, in a sense, always performing and were a constant reminder of what the former might become if their stage careers faltered: 'I asked . . . what drag queens do when they are out of work, he said, "They get their butts out on the street, my dear, and they sell their little twats for whatever they can get for them."'[221]

Drag was of the streets as well as the stage. This was the world of Rechy's *City of Night* that so scared the heterosexual cross-dressers:

I know the scene: Chuck the masculine cowboy and Miss Destiny the femme queen: making it from day to park to bar to day like all the others in that ratty world of downtown L.A. which I will make my own: the world of queens and malehustlers and what they thrive on, the queens being technically men but no one thinks of them that way – always 'she' – their 'husbands' being the masculine vagrants – fleetingly and often out of convenience sharing the queens' pads – never considering theyre involved with another man (the queen), and as long as the hustler goes only with queens – and with other men only for scoring (which is making or taking sexmoney, getting a meal, making a pad) – he is himself not considered 'queer' – he remains, in the vocabulary of that world, 'trade'.[222]

Rechy used the contrast of effeminacy to highlight his own, equally carefully crafted, masculinity. In *City of Night*'s celebration of the sexual dialectic of hustler/trade masculinity and fairy effeminacy, his sympathies and identification were clearly with the former, the 'tough-looking masculine hustlers', 'the handsome masculine ones desired alike by men and women'.[223] His 'femmequeens' are noisily defiant in their effeminacy.[224] But he was disparaging of these 'painted sallow-faced youngmen [*sic*], artificial manikin faces like masks', 'the gushing swishes, hands aflutter like wings', those 'prematurely sentenced to a purgatory of half-male, half-female'.[225] Though their dress and speech are described in great detail, the portrayals are never celebratory. For Rechy, these are caricatures of femininity, attempts to disguise a masculinity that invariably declares itself.[226] Their happiness is as false, as precarious, as their make-up; he refers to 'a franticness that only abysmal loneliness can produce'.[227] They are the 'mutilated sex', with 'woman-act[s] so exaggerated, so distorted'.[228] The drag queen Miss Destiny gets a whole chapter to herself, 'Miss Destiny: The Fabulous Wedding'. However, the fabulous wedding never happens, or rather becomes so mythical that it slips away. Miss Destiny came to Los Angeles hoping to make it as a star, 'that some producer would see me, think I was Real – Discover me! – make me a Big Star . . . and we would stand in the spotlights and no one would ever know I wasn't Real'. It was, the narrator intervenes in parenthesis and italics, *'That impossible strange something that will never happen'*.[229] Like the drag queen more generally, Miss Destiny's destiny was her unreality. Yet if we read through Rechy's prejudices, the crucial point is clear: the drag queen was an essential part of the American urban sexual landscape in the 1950s and 1960s.

This was the world too of Hubert Selby's powerful *Last Exit to Brooklyn* (1957), the hard-drinking, violent, criminal, blue-collar, woman-hating, bar-based environment, populated by workers, soldiers, sailors, and criminals, their women (girlfriends, wives), female sex workers, *and fairies*. The worker was as liable to fuck a fairy as a wife or a female sex worker. He was most likely to receive oral sex from the fairy: 'I am an expert in my field honey. No body [sic] can suck a cock better than I.'[230] Indeed, much of the sexual activity in the novel is between fairies and working-class toughs – referred to as 'rough trade' by the fairies. Homosexuality in *Last Exit* is expressed as effeminacy:

> Georgette was a hip queer. She (he) [sic] didnt try to disguise or conceal it with marriage and mans talk, satisfying her homosexuality with the keeping of a secret scrapbook of pictures of favourite male actors or athletes or . . . leering sidely while seeking protection behind a care- fully guided guise of virility . . . but, took a pride in being a homosexual by feeling intellectually and esthetically superior to those (especially women) who werent gay (look at all the great artists who were fairies!); and with the wearing of womens panties, lipstick, eye make-up (this including occasionally gold and silver – stardust – on the lids), long marcelled hair, manicured and polished fingernails, the wearing of womens clothes complete with padded bra, high heels and wig.[231]

It was the femininity of the fairy that was the lure for rough trade. 'Harrys eyes bugged when he saw Lee. She looked like one of the show girls you see in some of the magazines (her hair was shoulder length and golden blond and she was always smartly in drag), a real doll.'[232] Vinnie was 'hip to Lee, but she still looked like a lovely doll and he thought of her as a dame'.[233] Fairies were fucked much as women were fucked: 'comon motherfucka. You wanna look like a broad ya gonna get fucked like one . . . Hey Vinnie, come-on. Lets throw a hump intaer.'[234]

Those like Georgette in *Last Exit* are presented as female with a hinted maleness – as in the 'She (he)' of Georgette's introduction. They have female names (Georgette, Camille, Goldie, Lee, Ginger, Alberta, Regina), and are always referred to with feminine pronouns. It takes a page for Georgette's masculine origins to be established with an act of violence when her 'older brother slapped her across the face and told her that if he ever came home like that again hed [sic] kill him [sic]'.[235] But male same-sex desires are inscribed with the use of the terms queer, bisexual, fag, faggot, freakish, and (as in the quoted extract) reference to 'her homosexuality'. But it is a strange, a queer, homosexuality. Not hers, really, if she saw herself as a woman. Not that of her male partners, surely, who prided themselves in their heterosexual masculinity: Georgette and her like were effectively women. Harry attends a drag ball, attended by fairies and their johns and trade: 'He was surprised, though he knew they were men, how much they looked like women. Beautiful women. He had never in his life seen women look more beautiful than the queens strolling about the floor of the ballroom.'[236] But there are also lingering reminders of masculinity: in the scene where Harry threatens Georgette with his switchblade, 'I/ll [*sic*] makeya a real woman without goin ta Denmark'; when Ginger resists the urge to yell at the brutish Harry, 'IM MORE OF A MAN THAN YOU'; or when the same Harry awakens with Regina's cock against his mouth, 'looking at her cock and the hand around it, the manicured and redpolished nails'.[237]

Conclusion

What was then known as transvestism was surprisingly strong in its various forms throughout the so-called transsexual moment. Yet, as in *Last Exit*'s reference to 'goin ta Denmark', the transsexual alternative stalked cross-dressing. Significantly, Rae Bourbon cut an album called *Let Me Tell You About My Operation*, based on his claimed gender-affirming surgery in Mexico in 1956. This may have been a publicity stunt, inspired by the success of Jorgensen – indeed, upstaging her with the claim that Bourbon's was the first such operation in North America. But the significance is that it was a claimed shift from impersonation to real womanhood: 'Female Impersonator Needn't Fake Again – Says Surgery Made Him A Her', as the headline on the record cover expressed it.[238] On the face of it, Bourbon had made Benjamin's move from enacting the role of a woman to actually being one. Recall, too, that a few of the female impersonators in Seattle saw themselves as transsexual in the 1960s.[239] Tommye, who worked at the 82 Club throughout that decade, recalled a change in the late 1960s when 'they started the hormone injections and the silicone'.[240] And when Newton was researching her drag shows in the mid- to late 1960s, transsexuality, in the form of hormone use, breast implants, and surgery, was intruding onto the scene. Purists among the performers disapproved.[241]

Turnabout, we have also seen, catered to transsexual readers – not always positively (recall Benjamin's warning about anti-transsexual attitudes). *Transvestia* was generally hostile. Those associated with the magazine might use electrolysis and take hormones to enhance their femaleness, but they did not favour gender-affirming surgery. Prince argued that few of those seeking surgery were 'proper Transexuals [*sic*]'. Most were 'mistaken' (in ways that will be explored in a later chapter) and among the ill-informed were 'misguided TVs'.[242] Some key articles were aimed at dissuading the allegedly confused: 'Hormones and Surgery: Yes or No', and 'Should I?????' predictably concluded in the negative.[243] But the mere existence of these opinion pieces demonstrated a trend: 'I wonder if there is one, just one, single solitary transvestite who, at one time or another, didn't give serious thought to the possibility of a sex change.'[244]

Transsexuality lurks in the background of the transvestism described in Beigel's 'A Weekend in Alice's Wonderland': Alice was Susanna Valenti / Tito Valenti and the wonderland was Casa Susanna.[245] Katherine Cummings, who, like Beigel, had attended Casa Susanna and who had many cross-dressing friends in the 1960s, took the transsexual path.[246] She knew a Pennsylvania millionaire, Rex/Gloria, who, Cummings claimed, used their wealth to fund 'what she called her "girl factory"', paying for and supporting young trans women through their gender reconciliation.[247] Others – Prince and Valenti herself – decided, without surgery, to become what the latter referred to as full-time girls. Instead of occasional cross-dressing, they permanently cross-dressed – though, given their female identification, it really was not cross-dressing at all. Susanna referred to it in terms of living as a woman with a 'boy-within' rather than as a man with a 'girl-within'. She still distinguished herself from transsexuals, despite recognizing that they had 'many things in common'.[248]

It was inevitable, then, that some of the transvestites gravitated towards transsexuality. It was a time retrospectively depicted in the television drama series *Transparent*, as the previously transvestite Maura (then Mort) Pfefferman becomes convinced of her transsexuality – though the makers of the series transpose this historically specific moment to a time decades later.[249] Benjamin had recognized the possibility of a shift as early as 1953 – ironically, while drawing a distinction

between the two 'syndromes': 'It is quite evident that under the influence of sensational publicity a reasonably well adjusted transvestite
could become greatly disturbed and fascinated by ideas of surgical
conversion so that his [*sic*] emotional balance may be endangered.'[250]
The relationship between the two will be examined further in the following chapter, for the liaison was by no means at an end.

3

Blurring the Boundaries

Introduction

From its first real impact in the USA in the 1960s and 1970s, trans-sexuality exhibited some rather fixed definitional traits. The selves that were constructed by surgery and endocrinology were matched by what Douglas Mason-Schrock has termed 'new self-narratives'.[1] Indeed, during the early years of transsexuality's history, such narratives were required before radical surgery could be granted. 'The most dominant narrative form in these stories', wrote Darryl B. Hill, who has read many of the letters of young transsexuals to the therapist Harry Benjamin in the 1960s and 1970s, 'was the wrong body story'.[2] As an 18-year-old male wrote in 1970, after he had read about Christine Jorgenson, 'I have been very unhappy all through my life because my sexual organs are those of a male, but my soul, my inner being, is that of a female. I long to have my body match my inner being.'[3] Elliot Blackstone, a San Francisco Police Department community relations liaison officer in the Tenderloin in the 1960s and early 1970s, responsible for links with a large transsexual population, said that most of those he talked to 'picked up on this concept . . . "a woman trapped in the body of a man", or "a man trapped in the body of a woman" . . . That's how they started.'[4] 'The catch-phrase of transsexuals was that they felt like one sex trapped in the body of the other sex', Janice Irvine has explained: 'Sexologists waited for this phrase, or a facsimile of it. Soon they began hearing a master narrative repeated by surgical candidates who had studied the literature thoroughly and were prepared to jump the therapeutic hurdles set up by gender scientists.'[5] We will return to these self-narratives in due course.

When she was a teenager in the 1970s, Susan Stryker looked up 'Transsexualism' in the library to see if she could locate her own constellation of feelings:

> I feel like I'm really a girl so I *could* be transsexual, but if I'm transsexual I'm supposed to want to be with guys, but if I'm transsexual in order to be with guys then that means I'm repulsed by homosexuality, but I'm actually attracted to homosexuality, especially homosexuality in women, but a homosexual woman wouldn't like me because I have a guy body, but I could be homosexually involved with women if I were a woman, and I could be a woman if I was transsexual, but I can't be transsexual because that means I'm attracted to guys and repulsed by homosexuality . . .[6]

Similarly, Sandra Mesics, another trans woman, looked back to the 'thriving underground network of transsexual people' of the 1970s, sharing information about good and bad electrologists and surgeons, where to procure hormones, and, more to the point, 'what to tell the psychologists if you wanted to be approved for surgery'. 'You told them you hated your penis', she continued; 'You told them you liked men. You told them that you just wanted to blend in with society as a woman, get married, and settle down.' The problem was, Mesics confessed, that she did not really hate her penis, liked women, and was quite happy not blending in with society.[7]

As the recollections of Stryker and Mesics suggest, there is evidence that not all conformed to the rather restricting narratives of transsexuality, and that those who argue that this literature turned transsexuals 'into sexual stereotypes' have exaggerated this effect.[8] There was a blurring of boundaries between transsex, transvestism, and homosexuality. This chapter explores that realm.

Of course, we are all familiar with the imagery and sexual stories of postmodern sex. It is unremarkable that transgender in the 1980s, 1990s, 2000s, and 2010s brought such sexual malleability; it is discussed in Chapter 5. But the argument here is that we can detect a similar blurring of boundaries in an earlier period. Joanne Meyerowitz once observed of this period – the 1950s through to the 1970s – that the medical experts 'simultaneously defined the categories and undermined them', and that sexual and gender identities 'were neither entirely fixed nor entirely fluid'.[9] It is worth pursuing these insights more extensively.

Self-narratives

The stereotypical narratives are evident in transsexual autobiographies. 'I was three or perhaps four years old when I realized that I had been born into the wrong body, and should really be a girl. I remember the moment well, and it is the earliest memory of my life.' Thus begins the memoir of the writer Jan Morris, a trans woman.[10] Mario Martino, a trans man, writing in the late 1970s, referred to his pre-surgical condition as 'being cast in the wrong body . . . the imprisonment of body and soul'.[11] Mario always thought that he was a boy; or rather, 'Till I was nine, I didn't know that I wasn't a boy.'[12] Accordingly, photographs that appear to be normal pictures of a young female child become, in retrospect, clear visual evidence either of ambiguity ('evident at an early age') or of the 'true' male self ('A boy's face in banana curls').[13] Bernice Hausman has argued that the fact that such narratives may not correspond to the life experiences of many transsexuals is beside the point. Such texts established a 'discursive hegemony' copied by those who sought out gender reassignment.[14]

The notion of dissonance between gender and sex was invariably accompanied by formulaic conceptions of masculinity and femininity. Thomas Kando noted the reactionary social attitudes of some of the trans women he interviewed in Minneapolis in the 1960s, who endorsed 'such traditional values as heterosexuality, domestic roles for women, the double standard of sexual morality . . . They are the Uncle Tom's of the sexual revolution.'[15] One such woman told him that 'The ultimate criterion of being a woman is being a good wife, being able to make a man happy.'[16]

Morris certainly behaved in very stereotypical female ways once her transition from James was complete, becoming more emotional and less forceful. Instead of being 'thrusting, and muscular' (as James), she was yielding and accepting (as Jan). Her conversation lacked (manly) purpose and became 'meandering gossip'. All in all, a rather depressing précis of gender attitudes, which presumably were shared by both Jan and James.[17]

We can see the beginnings of this process in the 1960s in Benjamin's influential publication *The Transsexual Phenomenon* (1966) and Richard Green and John Money's edited collection *Transsexualism and Sex Reassignment* (1969). Here, would-be transsexuals had ready narrative models for the syndrome: 'I consider A. D. to be an excellent example of male transsexualism [what would now be termed the trans woman]', the author of one case study wrote, 'and her story is quite typical of that which the true transsexual presents'.[18] K. M., another assigned male at birth, had from childhood been 'more comfortable in a female role', had been cross-dressing since a young age, and was engaged sexually with men, though had clearly managed to persuade her consultant that she was living a chaste life latterly and that her motive for wanting sex reassignment was as a woman rather than to have sexual access to men. She already worked in a feminine occupation as a hair stylist, so demonstrated an ability to function work-wise: 'To summarize: K. M. is not only characteristic of male transsexuals [*sic*], but is considered an excellent candidate for sex-reassignment surgery.'[19]

Experts were aware that their case studies might prompt transsexual journeys: 'it is well known that a widely circulated article regarding a transsexual's dramatic transformation is often the precipitating event which prompts other transsexuals to accelerate the search for their transformation'.[20] If transsexuals read Green and Money – and Money thought that they 'read extensively about themselves' – they would also glean the necessary psychological profile of the gender identity required to justify reassignment surgery.[21] Their condition, they were informed, began at a very early age as what were termed 'pretrans-sexuals'. Femininity in 'males', whether effeminate homosexuals or male-to-female transsexuals, was the result of 'too much mother and too little father'.[22] 'It is no news that domineering, over-protective mothers and weak, passive, distant fathers are often the parents of effeminate men', wrote the gender expert Robert Stoller.[23]

Male-to-female transsexuals were also told the true/false questions asked on the masculinity–femininity scale: 'You would like to go
hunting with a rifle for wild game'; 'You can look at snakes without
shuddering'; 'You would rather study mathematics and science than
literature and music'; 'You would rather be a forest ranger than a dress
designer'; 'The sight of an unshaven man disgusts you'; 'When you
become emotional you come to the point of tears.'[24] They would, of
course, require a low masculinity score: no unshaven forest ranging or
wild-game-hunting for these candidates. Because clinicians considered
trans women particularly adept at mimicking female behaviour, they
needed to establish further evidence of their psychosexual condition,
eliciting an earlier gender history through interviewing. Again, the
literature provided clues. 'Invariably, the male transsexual [*sic*] patients
recalled an aversion to fighting, to boys' competitive games, and to
rough, outdoor activities. Labeled "sissy" by their peers, they much
preferred the security of the home and little girl activities.'[25]

Their reported sexual activity had to be equally gender-stereotyped.
Males were easily aroused, active, initiating. Females were slower to
arouse, passive, receptive. Money and his co-author Clay Primrose
were aware that such strict distinctions were frequently flaunted in
'everyday heterosexual relations'; 'In the case of the male transsexual
[*sic*], however, it would appear that his [*sic*] conception of sexually
dichotomous behavior is defined in conformity with the stereotypes,
not their violations.'[26] Hence, trans women played the receptive roles
in their physical relationships with men: they were the insertee in anal
or interfemoral intercourse, and the receiver of the penis in oral sex.
They felt either ambivalence or dislike for their own penis, the mark of
their incongruent maleness, and most had little sexual experience with
women. One married 'man' reported that 'he' could only achieve sexual
arousal with 'his' wife when 'he' imagined that she had the penis and
'he' had the vagina. And the men that they had sex with they considered
heterosexual, an opinion endorsed by their health professionals: 'Their
sexual relations were with male partners whom they considered to be
primarily heterosexual and erotically masculine – and justifiably so, if
we judge by the partners whom we have interviewed.'[27]

Female-to-male transsexuals could adapt the male-to-female model
but they were also provided with their own profile in a chapter by

Money and John G. Brennan, based on six Johns Hopkins cases.[28] They scored high on the M–F test, and were less feminine than 'normal women'.[29] When young, they were 'tomboys' (the keyword equivalent to the trans woman's 'sissy'), were energetic, fought, had male buddies, disliked female attire, cross-dressed, and generally did not play with dolls.[30] Their reported sexual activity was interesting because it was filtered through a double gender-stereotyping – for, clearly, the medical experts expected those assigned female at birth to observe 'reactivity . . . like that usually expected in the reactivity of the female to the male', while simultaneously demonstrating masculine sexual characteristics.[31] Hence, the trans man was 'romantic rather than genitopelvically sexual'.[32] When they did engage in sex it was with women and they were the penetrator (with finger or artificial penis) rather than the penetrated. All but one were living as males. Nearly all perceived themselves as male in appearance and were identified as such by others, and wanted (or had had) their breasts removed and a hysterectomy, and said that they aspired to male genitalia. Most reported 'intercourse imagery [with women] of the self as male', and that they fantasized, dreamed, and masturbated with a male self-image: 'patients saw an image of their bodies as male, without breasts and usually complete with penis'.[33]

In short, then, the selves that were reconciled with their bodies by surgery and endocrinology were matched by accompanying sexual and gender histories. Transsexuals were those who felt like one sex trapped in the body of the other sex, and formulaic notions of masculinity and femininity invariably accompanied this idea of dissonance between gender and sex. In this framework, moreover, transsexuals were different from transvestites and homosexuals; the distinction was part of their definition. And yet, as already intimated, the claimed sexual certainty of transsexuality masked a world of far more ambiguous alliances and practices.

Categories

We could start in the years of the initial taxonomic separation of transsexuals from transvestites and homosexuals, for a considerable conceptual blurring remained in the face of the urge to distinguish categories. Benjamin had indicated as much in the 1950s when he wrote of transvestism 'merging into' transsexuality, and claimed the homosexual inclinations of all 'trans-sexualists'.[34] A decade later, he still considered 'a sharp and scientific separation of the two syndromes . . . not possible', and while his Kinsey-like measurement of the transvestite and transsexual spectrum was supposed to provide a more systematic distinction between the categories, it arguably achieved the opposite with its seven-type, 'sex orientation scale': normal; pseudo transvestite; fetishistic transvestite; true transvestite; nonsurgical transsexual; moderate intensity true transsexual; high intensity true transsexual.[35]

Categories were pondered and discussed in Benjamin's correspondence with the Los Angeles surgeon Elmer Belt during the 1950s and 1960s. Thus, Belt wrote of one patient in 1958, 'This boy is unquestionably a transvestite. Whether or not he is a trans-sexual is an undecided question. He seems to be a very unstable individual who may not know his own mind in this regard.'[36] Benjamin advised of another, 'I believe she should be classified more as a transvestite than as a transsexualist. The question of operation may not come up in her case.'[37]

As an endocrinologist, of course, Benjamin had always sent his patients elsewhere for surgery – hence his initial contact with Belt – and his attitude seems to have been that, if his patients lived well-adjusted lives as women (presumably after hormone treatment – otherwise, they would not have been his patients), there was no reason to proceed to surgery. He wrote in 1958 that he had 'several patients . . . who live the lives of women with their original anatomy intact'.[38]

The Benjamin archives also contain minutes taken at transgender support group meetings in New York in the late 1960s. The trans men included those with female names and those with male names, both pre- and post-surgical, passing and non-passing, as well as their support group of friends, girlfriends, and 'wives'. The trans women included an equivalent range of cross-dressers and varied surgical modifications, including one attendee who made a dramatic impression: 'The first meeting we had, there was someone from Baltimore who tried to talk everyone out of the operation. Meanwhile she is the prettiest and most feminine of all and she's a friend of Spiro Agnew and this gives her a lot of pleasure that the Vice-President of the U.S. doesn't even know.'[39] The respective groups discussed sex and sexuality. One trans woman talked 'about how to go to bed with a man before operation; fool him, play with him; say you're a virgin or you're afraid, etc.'[40] Ray, a trans man, said he did not 'consider myself a homosexual [has a wife]. I was in gay life a long time. I'm very happy this way.'[41] A trans woman 'talked about the problem of his [*sic*] appearance. People stare and can't decide if he's [*sic*] a dyke or a fag.'[42]

We could also return to the earlier-discussed female-to-male cases in Green and Money because one of their intriguing aspects is that they seem less prescriptive than the male-to-females. While there was an assumed model of progress towards masculinity, there was variation in its expression. Some lived as males before any treatment, either hormonal or surgical. One trans man had hormone treatment but not surgery. Another had a hysterectomy but had not taken hormones. Few (it is not clear whether it was one or two) had obtained a phalloplasty (the creation of a penis). Their self-assessment before hormonal intervention was also at variance to the stereotypical one-sex-trapped-inside-the-body-of-the-other scenario. One said that he was 'both sexes in one body'. One thought that he was sometimes male and sometimes female. And another 'recognized she [*sic*] appeared as a female (she [*sic*] was an exceptionally attractive appearing lesbian) with male attitudes'.[43]

Benjamin's *Transsexual Phenomenon* contains the autobiography of Joe, a female-to-male transsexual who, when he wrote the preface to his brief life account in 1965, had successfully transitioned after hormone treatment, breast reduction, and a hysterectomy, and was legally married to a woman.[44] But the autobiography itself was written in 1956 at the moment Joe found a sympathetic doctor and discovered from a male-to-female transsexual friend that 'the sex change possible for males might also have its counterpart for the female'.[45] The account is interesting in that it covers the history of someone who desired masculinity in what was, for them, a time before transsex – until, as he put it, 'science . . . carved out this niche for me'.[46] The narrative is of a person who thought that they had 'all the traits of a boy but not the right physical attributes', who was not attracted to males, except as friends, had minimal heterosexual contact (although those fleeting moments involved two marriages, and two pregnancies and miscarriages), and who had difficulty making sense of their predicament.[47] 'Homosexuality' was the only framework available as they charted close female friendships, woman–woman sex, and a longer-term relationship – described as 'homosexual' – where Joe was the cross-dressed 'husband' living with a 'wife'.[48] Though he repeatedly classified himself as homosexual, and described his sexual practices (as a woman) with women, including 'dyking', Joe did not think that he quite belonged, 'that I was in quite the same category'.[49]

L. M. Lothstein's 1983 study of female-to-male transsexuality, of lives lived in the 1970s and earlier, was dedicated to demonstrating that 'female transsexualism [those we would now term trans men] is not a unitary phenomenon', and he seemed to take delight in challenging assumptions at every turn.[50] Patients may have invoked the stereotypical 'I'm a man trapped inside a woman's body. I'm in desperate need to match my body with my mind, the real me inside', but there was a chasm between the presented certainty and the social–sexual flux of individual lives.[51]

Hence ,'Barbara/Brian' lived and worked as a male and had a 'wife' and two children. To his neighbours, claimed Lothstein, he was a respectable, church-going family man. But his childhood and youth had been violent and promiscuous, and he had engaged in both heterosexual and homosexual sex. One of his children was arranged to provide a previous female lover with a child; the other was the result of a homophobic rape. His wife was the ex-lover of a trans man who left him for Brian's partner – in effect, swapping lovers. Before Brian identified as transsexual, he was a lesbian, with a procession of short and longer-term, 'same-sex' relationships. The case file reports a wild four years when 'Barbara engaged in a series of frenetic, tandem gay relationships. She [*sic*] was profoundly alcoholic – going on binges and suffering from delirium tremens – and while identifying as a lesbian, wore male clothes and alternately played a male/female role.'[52]

Then there was 'Randi/Randy', whose transsexuality was doubted both by his older brother (seemingly well informed on such matters) and by Lothstein, whom he had contacted for help. The brother was not hostile to sexual unorthodoxy; he was willing to accept what he saw as Randy's lesbianism or bisexuality but questioned the authenticity of 'her' quest to become male. Lothstein's enquiries into his patient's case history revealed that the transsexuality emerged in their teenage years, after running away from home and links with gay circles: 'Within a short period of time she [*sic*] was immersed in the gay and transsexual culture, doing a male impersonator act and dating other women in the guise of a man; and living with two male-to-female transsexuals who befriended her and facilitated her role transformation to male.'[53] He was convinced that two periods of such influence – in Europe and then later in the USA – had led towards transsexuality (Randy may have argued that such environments provided the space for personal identity to develop). The therapist hypothesized that once Randy moved in transsexual circles, the 'life style' provided a 'solution to her [sic] bisexuality' and 'an explanation for her [sic] confused inner feelings, and the basis for an identity'.[54]

His next case study was 'Patricia/Pat', 'the crazy kid next door', who had suffered a terrible young life of violence and sexual abuse, had lived as a sex worker, had engaged in both heterosexual and lesbian sex ('bisexual experiences with just about anyone'), and seemed to grasp at masculinity as a way to solve 'her' problems.[55] Pat had unrealistic expectations of surgery, wanting to 'masculinize herself [sic] so that she [sic] looked just like her [sic] second husband, Ken'. 'Patricia asked if we could make her [sic] 6' 2" and 185 pounds (she [sic] was now 5' 4" and weighed about 100 pounds).'[56]

Lothstein saw such cases as examples of the multifariousness of transsexuality, evidence that it was 'a complex phenomenon which was multiply determined'.[57] Barbara/Brian's sexually chaotic family life undermined any possibility of a 'core female gender identity': 'Her [sic] experience of a continual assault upon her [sic] normal female gender development led to profound gender diffusion and dysphoria throughout her [sic] life cycle.'[58] Randi/Randy lacked 'a cohesive self-system', constantly 'searching for an explanation and resolution of her [sic] inner confusion'.[59] Patricia/Pat's 'two marriages to men, sexual preferences for women, and bisexual gender schema, suggested a confused, chaotic self and gender identity'.[60]

Lothstein also cited an earlier case study of 'Gloria/Ray' who believed himself to be male and acted accordingly, adopting, in Lothstein's estimation, an exaggerated, stereotypical masculinity. Although Ray saw his 'true self' residing in this masculinity and his 'false self' reflected in his 'sexual anatomy', and claimed to be secure in his identity, Lothstein was more sceptical, pointing to interactions where this certainty must have been challenged.[61] This was especially the case with his strange relationship with a married woman, Virginia. Ray lived with Virginia and her husband and children, but did so as Gloria so that they could carry on a sexual relationship 'without arousing her husband's suspicions'.[62] When they had sex, Virginia called Ray by his male name, indeed called out 'Ray, Ray' at the point of orgasm, but neither Ray nor his therapist were convinced that Virginia was not responding to Ray as a woman, that is engaging in lesbian sex, even when calling him Ray.[63]

> Ray's statement that 'he' has no confusion about 'his' sexual and gender identity (a claim made by most female transsexuals) contrasts with the known facts in which multiple sexual and gender disturbances are manifested as Ray moves in and out of the interpersonal underworld and is subject not only to 'his' personal representations of reality but the confirmation or nonconfirmation of that perception by significant others.[64]

These were not simple cases of women achieving the opposite-sex identity that they had craved from their first days of self-awareness.

From Gay to Trans

It is noticeable in the personal histories of trans men that many had spent portions of their lives as lesbians. In a sense, this is unsurprising. The butch role in twentieth-century lesbianism, which Elizabeth Lapovsky Kennedy and Madeline Davis have argued governed both personal behaviour and community structure among the women they interviewed, was almost a prescription for the masculinity of trans men.[65] The remembered life histories of 1950s and 1960s butches in Buffalo city prefigure trans childhoods: playing like boys, dressing in boys' clothing. 'I always felt that I was in drag in women's clothing even as a child', could have been a trans man speaking.[66]

Jamison Green was lesbian in the 1970s long before he transitioned to manhood and joined a men's group.[67] That was how he made sense of his feelings at the time. It was not until the late 1980s that he began his transsexual quest. Green knew of others. Loren Cameron, the photographer who has visually charted his masculinity without a penis, was a lesbian for more than ten years.[68] Of the forty-five FTM (female-to-male) transsexuals interviewed by Holly (later Aaron) Devor in the late 1980s, over a third had considered themselves to be lesbian before they discovered female-to-male transsexuality, and another third engaged in same-sex relations while resisting the identity of lesbianism. Devor claimed that 80 per cent of his subjects 'were attracted to the possibilities offered by lesbian roles', even if the attraction was temporary: 'The combination of their own alienation from being women-identified and their partners' insistence upon relating to them as women and as females prevented participants from adopting enduring identities as lesbian women.'[69] Thirteen of Henry Rubin's twenty-two 1990s FTM interviewees were former lesbians.[70] Whether we can go as far as Rubin in positing a direct historical link between a 'lesbian-feminist revolution' and the increasing visibility of the trans man in the 1970s is another matter, but it is certainly an intriguing hypothesis.[71]

But can a similar argument be made for male homosexuality? There are hints. Early trans women were seen as homosexuals. Even after the technology of gender reconstruction was available, some medical experts considered such people to be homosexuals in severe self-denial – the logic being that they thought that the only way that they could gain guiltless sexual access to men was by becoming women.[72] That male-to-female transsexuals were basically homosexual was an opinion voiced in a counsellors' guide prepared by experts for the Erickson Educational Foundation in the 1970s – an organization extremely sympathetic to transgender.[73] At a more popular level – if *Take My Tool* (1968), pulp fiction masquerading as memoir, is any guide – the male-to-female transsexual could easily be seen merely as a homosexual 'male' becoming a woman ('As time went by after adjusting my life from a homosexual to a woman'). With encounters described in lurid detail, the object of desire remained the same: the masculine, heterosexual male.[74]

Arthur James Morgan of the University of Pennsylvania's Hospital calculated that 30 per cent of his patients in the 1970s were homophobic homosexuals. 'It is hard to believe that in the last quarter of the 20th century in a large and sophisticated urban area there is a sizable group of homosexual men [*sic*] who would prefer (and whose families would prefer) penile amputation and castration to coming to terms with their homosexuality.'[75]

The male-to-female transsexuals of the late 1960s are particularly interesting because they seem to have spanned the time when transsexuality became a possible sexual identity for those who had hitherto identified as effeminate homosexuals. When Brenda Dott (formerly the drag queen Bobby) performed as the '$10,000 Woman' in a white, working-class, Pittsburgh, gay social club in 1976, revealing the results of her recent gender reassignment surgery, it was but a later and more dramatic illustration of a social shift.[76] The rock singer Jayne County, who considered herself transsexual in the 1970s, said of the 1960s that, regardless of whether one was a 'screaming street queen, or a full-time drag queen . . . you were gay . . . That's what we were: flaming creatures, just like the title of the Jack Smith film.'[77] In fact, the male effeminacy of that period beckons further study, worlds like that of the future Cockette and disco star Sylvester in Los Angeles in the mid-1960s, where young gay boys, the Disquotays, donned women's clothing, wigs, make-up, padded their bodies or took hormones, to 'become the most fabulous girls around': 'Folies Bergère in the ghetto', as one of them recalled to the sociologist Joshua Gamson.[78]

Two 1960s documentaries depicted this significant era. The gay drag queens in *The Queen* (1968) – at least those who expressed any opinion about gender-confirming surgery – were against it:

I'm proud of what I've got and I certainly wouldn't want it whacked off . . . I have enough money to go through the sex change and I live only thirty miles from Johns Hopkins but it is the last thing I would want. I know that I'm a drag queen . . . been gay for a long time. But I certainly do not want to become a girl.

'Even if I could have a sex change I wouldn't have it anyhow . . . no, my goodness gracious no.'[79] They were performers, and part of their appeal was their ability to transform their masculinity temporarily on stage: the documentary recorded the drama involved in a national drag queen competition held in New York in 1967. So, while the two men filmed in discussion were uninterested in gender reassignment, it is significant that it was an issue in the first place, and that they were talking about it as a possibility. It should be noted that the winner of the competition,

Rachel Harlow (Richard Finocchio), did in fact undergo such surgery in 1972.[80]

If *The Queen* was slanted towards effeminate men ('you really should have been a girl', one was told by the draft board) who were comfortable with their homosexuality, *Queens at Heart* (1967) featured what would later be termed male-to-female transsexuals (though actually the term 'transsexual' was never used in the film). And if *The Queen* was an insider documentary, *Queens at Heart* was firmly outsider in perspective and framing, though its subjects, four women ('who were really men'), tried valiantly to preserve their dignity: 'Lets talk for just a moment about your sex life'; 'It should be obvious to you [pause] that you don't lead a normal life.' These 'four contestants in a recent beauty contest', Misty, Vicky, Sonya, and Simone, were also involved in the drag queen scene and were consistently proclaimed to be 'Homosexual' (with vocal emphasis on the *mo*) – 'Now Simone you openly admit that you are a Homosexual' – but, unlike the subjects of *The Queen*, had decided on a 'sex change'. All were taking female hormones and wanted gender-affirming surgery: '"Are you happy?" "Oh yes, I am." "What would make you even more happy?" "If I would be able to have a change from man to woman."' Vicky, who was engaged to a straight man, a photographer, said that 'Most of the homosexuals and drag queens they get to a certain age and they don't want to live no more, so I've just about hit that age . . . [talks of her own suicidal thoughts] . . . I just think about changing to a woman, hoping that would be an escape.'[81] They were, it should be noted, rather vague about what surgery actually involved.

The performers in these documentaries could have been described as gay cross-dressers, but they took different paths. Harlow and Vicky and the others wanted to transition, and the pair in *The Queen* resisted any such change ('no, my goodness gracious no').[82]

The trans activist Sylvia Rivera looked back on that period when she was a street queen sex worker on 42nd Street. She recalled that 'so many in the late '60s and early '70s ran up to the chop shop up at Yonkers General'. She had contemplated gender-reconciliation surgery herself but had decided against it: 'I feel comfortable being who I am . . . I always like to be an individual. In the beginning I decided that not getting the operation was because I wanted to keep the "baby's arm".'[83]

Members of the San Francisco-based Gay and Lesbian Historical

Society of Northern California once observed that transsexuality introduced a new classificatory hegemony into 'previously more heterogeneous transgender populations'.[84] James Driscoll, who was writing in 1971, said that 'almost all the transsexuals' whom he encountered in San Francisco's Tenderloin area 'reported that they considered themselves to be homosexuals in late adolescence'.[85] They moved with ease in San Francisco's gay world, he argued, continuing quickly – in a matter of months, rather than years – through various sexual identities. First they were effeminate homosexuals, 'hair fairies', with long hair, 'a little make-up', loose sweaters, and tight trousers.[86] Then, after contact with a 'transvestite subculture', they became homosexual transvestites, 'coming out' (their own language) as self-designated 'queens' or 'drag queens'.[87] Finally, when they acquired knowledge about transsexuality and the possibility of bodily modification, they were able to conceive of themselves as transsexual rather than homosexual or transvestite: 'Once they have heard of the conversion operation and know that there is such a thing as a transsexual, the self-concepts of the girls seem to change. Now they regard themselves as women in every sense of the word except one. This female identification is very strong in the transsexuals and dominates all other aspects of their lives.'[88] Elliot Blackstone, a sympathetic local policeman, said that there had been 'no transsexuals' in the Tenderloin in 1967, but, by 1977, 'there were three thousand'.[89]

It is intriguing that the shift from transvestism to transsexual in San Francisco in the 1960s was almost a mirror image of a similar, slightly earlier, phenomenon in Paris in the 1950s and 1960s, when the performers in cabaret there (including the famous Coccinelle) discovered hormones and then availed themselves of the services of Georges Burou in Casablanca, the pioneering gender-reconstruction surgeon discussed previously in the book, and which Maxime Foerster sees as a formative stage in the history of French transsexuality.[90] Coccinelle herself referred to breasts 'sprouting like mushrooms' on the chests of the female impersonators, the *travesti*, in Paris at that time.[91] April Ashley performed there and likewise visited Burou; there are harrowing descriptions of her surgery.[92] She said that she thought that all the girls at Le Carrousel were related because they had identical noses, only to be told that that was because they all went to the same plastic surgeon.[93] Ashley was aware of the varied character of the female impersonators:

some had no desire to be women, some thought they were. Ashley numbered herself among the latter; 'Me? I just wanted to be a proper girl.'[94] Peki d'Oslo, later known as Amanda Lear, was another transsexual member of the troupe. She would later become friends with Salvador Dali and the lover of the Rolling Stone Brian Jones, Roxy Music's Bryan Ferry, and David Bowie.[95] (She appears on the cover of a Roxy Music album.) This world of trans performers and sex workers is portrayed beautifully in the photographs of Christer Strömholm, who lived among these women from 1959 into the 1960s.[96]

Many of the San Francisco trans women survived by hustling, by sex work. Blackstone, who knew 'literally hundreds' of these women ('my gals'), said that 'the only way they could make a decent living was [by] hooking'.[97] Driscoll, the author of the Tenderloin study, lived in a hotel in San Francisco in the late 1960s that was totally occupied by trans women and their 'husbands': 'Every one of the girls in my sample was a prostitute. In fact, the hotel we shared was nothing but a whore house.'[98] Whereas previously they had been effeminate, homosexual, transvestite hustlers, they were now women engaged in sex with men:

> Prior to this phase they considered themselves to be homosexuals. Now they deny any such status and claim that they are normally sexed. When asked how they account for this when they admit to performing homosexual acts, the girls have told me that they think of themselves as women. This identification is so strong that to them it would be a perversion to choose a female sex partner.[99]

An article in San Francisco's newspaper *Good Times* in 1970 outlined the case of Lizzie, one of the many 'drag queens' who hustled in the Tenderloin:

> She has been working in the Tenderloin seven years and wants to stay because it is the only place she knows where transsexuals can be accepted. Lizzie has lived in almost every hotel-apt. in the area and has been busted over 100 times ... Lizzie is taking female hormone pills, receiving silicone injections and wants eventually to undergo a sex change ... Originally from Texas, Lizzie became a sissy with flaming red hair and a high screeching voice when she arrived in

the Tenderloin. Now more feminine oriented, she complains of the depressions and loneliness of women. 'If I had a pussy I'd be a normal woman' . . . Lizzie doesn't like to get head but loves to give it and have anal intercourse . . . she is emphatically not a lesbian and could never make love to someone she feels is the same sex.[100]

This was indeed the case of 'a Tenderloin queen moving between different identity categories'.[101] This was around the time when 'Tenderloin Transsexual', a member of Vanguard – an organization of queer street youth from the Tenderloin, comprised of these very sex workers – wrote that her gender ambiguity was at an end: 'Not long ago I didn't know who I was . . . Now I know.'[102] She had shifted from homosexuality to transsexuality.

There must have been pockets of similar gender and sexual configurations throughout urban America, now all but lost to the historical record. A yellowing, cheaply printed pamphlet from the Washington DC Project of the Arts in 1976, a product of an exhibition called 'Another Washington', describes the environment of the New York Avenue / 13th Street neighbourhood, with its male and trans sex workers, where drag queens were being influenced by notions of sex-reassignment surgery.[103] The author of the piece, Joel E. Siegel, described the bar Dolly's (formerly The Famous), where all the waitresses and bartenders were 'transvestites and drag queens'. 'Dolly's isn't exclusively a gay bar, or a drag bar, or a neighbourhood straight bar', he wrote, 'but a combination of all these things whose proportions change with the evening's chemistry'.[104]

Siegel talked to some of them. Billie Jo, who was a waitress and former go-go dancer, said that she had not heard of drag queens before she came to Washington and started cross-dressing and taking female hormones. She held out hope for eventual surgery and was in a relationship with a man she called her 'husband', which she called a 'gay relationship', yet she seemed unsure of whether she should be described as gay.[105] Bobbie, a dancer at Dolly's, had been castrated and was taking silicone shots. She had worked stealth (that is not declaring her transness), and claimed that not even her agent knew: 'I've worked as a female stripper on the Midwest Circuit'; 'I'm so real now, I think I'm a lesbian. I've been dating a lesbian in fact.' She dreamed of going to Colorado for surgery, yet simultaneously realized that the costs involved were prohibitive – 'if I have to save for it, I'll probably never get it. Maybe I don't want it that bad after all.'[106] Finally, there was Liza, who tended bar, had worked as a sex worker on the Avenue, and still had regular clients at Dolly's to supplement her income: 'I have regular Johns'; 'They have it in their mind that you're a woman and, as long as you look like a woman, fine. Even if they're blowing you.' She was having electrolysis, had had facial silicone injections, and, like Bobbie, talked of going to Colorado: 'All the other queens I have met want to have sex changes to make money off of it.' As with the others, the (im)possibility of transsexuality (and the word was never used) meant transition from a gay identity; 'Gay life is cruel . . . In any case, I'll be happy being a woman and that's entirely different from gay life.'[107]

Flaming Creatures

The 1960s and early 1970s drag queens of experimental theatre and film in New York and San Francisco were part of this indecisive world. Stefan Brecht, who knew this milieu intimately, wrote in 1968 that the drag queen 'poses the problem of psycho-sexual identity: to what extent male and female conduct, masculinity and femininity, are social role-identities, cultural artifacts, what they are, might be, should be – how valid these roles are, how natural'.[108] In New York, there was Jack Smith's banned film *Flaming Creatures* (1963), described by Susan Sontag as poetic transvestism, where it was difficult to tell which actors were male and which were female: 'These are "creatures", flaming out in intersexual, polymorphous joy.' 'The film', she continued, 'is built out of a complex of ambiguities and ambivalences, whose primary image is the confusion of male and female flesh.'[109] The playwright and Andy Warhol collaborator Ronald Tavel claimed that reviewers of the film argued over whether one of the dancers, Mario Montez (René Rivera), then known as Dolores Flores, 'were male or female'.[110]

The 'drag queen is all over our year', Tavel wrote in 1965; drag was 'the symbol of our time'.[111] The occasion for his observation was his essay on the making of Warhol's *Harlot* (1964), in which Jack Smith's star Montez featured as Jean Harlow, suggestively eating a series of bananas, 'a near orgy of banana consumption' as Tavel put it.[112] But his remark about drag as a symbol of the time was equally true of a cluster of Warhol films and stars; not just Montez, whose Warhol portfolio is extensive, but Candy Darling, Holly Woodlawn, and Jackie Curtis, who appeared in the Andy Warhol / Paul Morrissey films *Flesh* (1968), *Trash* (1970), and *Women in Revolt* (1971).[113] See Illustration 13. All played memorable roles in the films, whether it was Jackie Curtis's visual admiration of the hustler Joe Dallesandro and Curtis and Candy Darling's conversation about the Hollywood movie magazines in *Flesh*, or the trio's women's liberation histrionics in *Women in Revolt* ('Warhol's women who are men' as *Vogue* billed it) – though to describe Woodlawn's bottle masturbation scene in *Trash* as an on-screen validation of the transsexual would be going a step too far.[114]

They were sometimes described as drag queens – Mary Harron called Candy Darling the 'Marilyn Monroe of drag queens' – but it is by no means certain that that was how Warhol's acolytes saw themselves.[115] Candy Darling sometimes pondered her identity, in snippets, most of them frustratingly undated. 'I tried to explain my identity as being a male who has assumed the attitudes and somewhat the emotions of a female', she wrote in a letter to a friend, probably in 1970.[116] One can almost hear her talk in some of her diary entries: 'I am not a genuine woman but I am not interested in genuineness. I'm interested in the product of being a woman and how qualified I am.'[117] 'My goal', she confided to her journal, 'is to be a beautiful woman, rich and married by 1971'.[118] She reminded her cousin of the movie magazines that they consumed while in their teens, revelling in the fact that she had just appeared in their favourite, *Photoplay*, as well as in *Vogue* and *Esquire*.[119] Around the same time, she declared that she had 'decided to be sex changed [*sic*]. I am too female to be half & half.'[120] But in 1972, she was saying that she did not want to be a woman anymore.[121] It is unclear whether, apart from hormone shots and electrolysis, she took any other steps along the lines of gender affirmation. When she was photographed by Richard Avedon as part of Warhol's Factory in 1969, in a group portrait now in Britain's Tate, Darling posed bravely in the nude, with her breasts partially covered by her hair but her penis fully visible. The Tate summary calls her 'a transsexual who appeared in Warhol's films', though trans woman would be a more appropriate description.[122]

13   *Jackie Curtis and Holly Woodlawn at The New York Cultural Center, 1974.*

Like Darling, Holly Woodlawn – Puerto Rican like Montez – lived her life as a woman most of the time, besides a momentary reversion to manhood and a (non-lesbian) relationship with a woman.[123] Woodlawn visited none other than Harry Benjamin for hormone therapy.[124] She also contemplated a 'sex change', but said that she had no idea of what was involved and was mainly doing it because she thought it would bring her closer to her partner.[125] She also claimed that she 'never once felt like a woman trapped in a man's body. I felt more like a man trapped in high heels.'[126] Her partner gave her the funds to go to Baltimore for the surgery, but she discovered (this was in 1966) that she would have to wait for a year and undergo psychiatric analysis, so she spent the money.[127] She described herself at that time as 'a woman regardless of what I was wearing': 'there are those men such as myself, who want to live as women and go to the extreme of shooting hormones and undergoing electrolysis treatments so they can look real'.[128] Woodlawn and her friends would dance in bars and have sex with straight guys who thought they were straight women. They would say that they were saving themselves for marriage, but were not averse to providing a blow-job. She was struck by the gullibility of her sexual partners.[129] When she appeared on the Geraldo Rivera show, she denied that she had always wanted to be a woman, and replied, when asked what it was like to be a woman trapped in a man's body, 'I'm not trapped in a man's body, I'm trapped in New York.'[130] Gender and sexual boundaries were somewhat hazy.

Woodlawn claimed that Jackie Curtis said that 'he never really wanted to be a woman'.[131] He certainly seemed to cross-dress for per-formance, rather than as a reflection of any true gender. Giulia Palladini has argued that his drag persona, which involved both male and female components, was a conscious effort to establish his place in the New York underground.[132] His photographic portfolio has pictures of him as a glamorous woman as well as a topless, hirsute man.[133] But it was a peculiar blend of masculinity and femininity that he made his own, with the imperfect make-up, five o'clock shadow, and op-shop clothes – what Palladini terms 'the imprecision of his drag outfit' – in total con-trast to the studied perfection of Darling.[134] His friends referred to his 'ratty house dresses, snagged nylons, and unruly hair', 'modeling her own deranged sense of vogue', as Woodlawn expressed it.[135] It is surely

significant that commentators, including Warhol himself, usually use
the description 'he' when referring to Curtis, and even Woodlawn
alternates between the masculine and feminine.[136] On the occasion
of his play, *Heaven Grand in Amber Orbit*, performed in 1969 by The
Playhouse of the Ridiculous (under the direction of John Vaccaro),
Curtis said she 'was not a boy, not a girl, not a faggot, not a drag
queen, not a transsexual – just me, Jackie'.[137] The interviewer, Rosalyn
Regelson, described Curtis's 'mini-skirt, ripped black tights, clunky
heels, chestnut curls, no falsies ("I'm not trying to pass as a woman"),
[and] Isadora scarf'.[138] It was a studied gender mix. 'She seems', wrote
Regelson, 'to have effected in her androgynous person, single-handed,
a social revolution'. And she quoted Curtis again: 'sex changes' were 'so
1950's'; 'I have to laugh at those people who say they feel like a woman
trapped in a man's body. What is a man? What is a woman?'[139] Curtis
reputedly told Warhol that it was 'easier to be a weird girl than a weird
guy'.[140]

The elements of transness varied among these stars. Despite Tavel's
reference to critics not knowing whether Mario was male or female,
Montez's masculinity was as evident as his femininity. Tavel's article
refers to Mario's wig, beard stubble, razor, shaving cream, and struggle
with his girdle and stockings, and then, after the shoot, his de-robing:
'Mario hastened to disdrag.'[141] Tavel said that he frequently failed to
recognize Montez in the street; he 'affected a walk more masculine than
a diesel'.[142] Significantly, when he performed a cameo in *The Queen*,
singing 'Diamonds Are a Girl's Best Friend', he was introduced as
'Mister Mario Montez'.[143] Indeed, there are excruciating moments in
both Warhol's prose ('I'd zoomed in and gotten a close-up of his arm
with all the thick, dark masculine hair and veins showing') and in his
film *Screen Test No. 2* (at the hands of Tavel, no less), where Mario's
masculinity is revealed: 'Now, Miss Montez, will you lift up your skirt?
... And unzipper your fly.'[144] Warhol distinguished between Montez
as a classic drag queen, only dressing as a woman for performance,
and what he called the 'social-sexual' cross-dressers (like Darling and
Woodlawn) who wore women's clothing more permanently.[145] See
Illustration 14.

Warhol would tire of his trans subjects. Drag queens 'were out',
he told the Italian art dealer who commissioned his series *Ladies and*

14 *Mario Montez in Andy Warhol and Paul Morrissey's* Chelsea Girls, *1966.*

Gentlemen in 1974. 'Candy was dead, and Jackie and Holly would drive him crazy, asking for more money every time they heard one had sold', so, according to Bob Colacello, one of Warhol's associates, they hit instead on a portfolio not of 'beautiful transvestites who could pass for women' but, rather, 'funny looking ones, with heavy beards, who were obviously trying to pass'. Warhol said, according to Colacello, 'We should get those nutty-looking drag queens from that awful place we went to that night . . . The one with all the blacks and Puerto Ricans, where I was robbed.'[146] Warhol's prejudices may have been revealing, but were ultimately eclipsed. The beauty, dignity, and exuberance of the *Ladies and Gentlemen* portraits undermine any purported ill-intentions of their creator.[147] As Jonathan Flatley has pointed out, it was one of the biggest series of paintings that Warhol produced, and the artist's 'excitement was evident in the energy he gave to the project'.[148]

With the Cockettes in San Francisco in the early 1970s, both act and audience proclaimed and subverted expectations and conventions of gender and sex:[149]

> Sequined head-dresses towering over mustachioed faces on satin and lace-swathed bodies of unrevealed sex, and in some faces a magnetic beauty could be felt . . . The audience was composed of members of every sex in every combination . . . A hermaphrodite with a mustache contradicting the evidence revealed by a fishnet shirt/blouse stalked coolly past . . . The acts are concocted of burlesque, majesty, insanity and sexuality . . . Couples woo with classic stratagems until the man is revealed to be a woman, without deterring the lovers.[150]

The Cockettes are especially interesting because it has been claimed that the word 'genderfuck' was coined to describe them.[151] Their historian and one-time member Martin Worman explained that they were hard to categorize. In terms of performing gender,

> the Cockettes were not just drag queens, nor could they be simply labeled as a gay theater company. Unlike traditional male or female impersonators, no attempt was made by the Cockettes to pass as another sex. For the Cockettes, drag was a generic term for whatever was being worn on or off-stage, whether for performance or going to

the grocery store . . . The Cockettes called themselves drag queens, but one would be hard-pressed to mistake a dolled-up male Cockette for a real woman.[152]

Jennifer Le Zotte has referred to '"glitter boys" with sparkling beards, hairy legs under miniskirts, or gowns torn to artfully frame penises'.[153]

What Worman said about real women was not exactly true of all Cockettes; Pristine Condition certainly paraded a feminine persona in her 1976 Tri-sexual Bicentennial Universal Calendar.[154] But Worman was correct about the juxtapositions found in individual bodies. Another image of Pristine Condition, in a pink, thrift-shop dress, matching high heels, and heavily applied make-up, had the caption: 'As she hitched up his dress, s/he said . . .'. As Julia Bryan-Wilson puts it, 'It was as if no one singular pronoun could encapsulate the ever-morphing identities Prissy is taking up, playing with – or perhaps *putting on*.'[155] The Cockettes' dress, their homemade costumes and thrift-store clothing, were central to their genderfucking, as they literally crafted selves.[156] See Illustration 15.

The critics recognized that this was a transvestism unlike other transvestisms, treating it first as a kind of innocent, youthful exuberance, and then, as things rapidly soured, as the death of the art. 'This is a drag show to end all drag shows', wrote the *New York Times* reviewer, 'the kind of exhibition that murders camp and gives transvestism a bad name'.[157] The group itself said that they presented 'sexual role confusion'.[158]

The Cockettes were equally confounding in their staged, indeed lived, sexualities. Hence, presumably, the 'Tri-sexual' calendar. As the former insider put it:

> To describe the Cockettes as a gay theater company is similarly restricting. The Cockettes were predominantly gay men, but not exclusively. Most of the women in the company and a few of the men were heterosexual, although most of them had occasional same-sex affairs; and many of the gay men had affairs with women. Indeed, the most flamboyant queens were more likely to do so than the butch-er gay men who were almost all exclusively homosexual. More importantly, the Cockettes did not present shows that were about gay life,

15　*Fayette Hauser, 'The Cockettes in a Field of Lavender, Marshall Olds, Bobby Cameron, Pristine Condition, 1971'.*

per se. It is true that their extravagant spectacles were imbued with a gay sensibility, an aesthetic which leveled myths of glamor and hetero-sexual romance to the ordinary, and elevated the ironic and perverse to high art; but this element had an appeal that went beyond a strictly gay audience.[159]

Again, the critics knew that something was going on. Rex Reed noted the strange innocence of their 'homosexual, bisexual, asexual and quad-risexual parodies of women's lib, old movies, politics and unisex that you'd never find in the more professional transvestite revues'.[160]

Significantly, in terms of the gender and sexual fluidity that I have argued marked trans history, a couple of the Cockettes and their circle transitioned to transsexuality. Bobby Cameron had sur-gery and worked as a showgirl.[161] Reuben, the young lover of the Cockette Dolores Deluxe, became 'a fully-blown woman named Ruby'; 'And I've seen him [sic] go through his transition. He [sic] was even my lover. We were all bisexual in those days too ... And you must have been really bisexual when you started being lovers with a pre-op transsexual.'[162] Bambi Lake, who moved on the fringe of the Cockettes and was associated with their offshoot group the Angels of Light, described herself as a 'pre-op transsexual'.[163] She began to take hormones in the early 1970s, she said, because of the access a femin-ized body gave her to straight men, and claimed that it was mainly the cost that prevented gender surgery, though hinted too that retaining her penis had its attractions.[164] But the majority of the Group were not transsexual.

Jimmy Camicia's 1970s New York, cross-dressing, gay theatre troupe, the Hot Peaches, drew on the movements that I have been discussing, for Camicia was inspired by the Cockettes and the Angels of Light. Curtis and Montez were involved in early productions, and Marsha P. Johnson, a trans activist whom we will encounter later, made regular appearances, including the role of chorus girl. Another of the Hot Peaches performers was Wilhelmina Ross, who (along with Johnson) appeared in Warhol's *Ladies and Gentlemen* (1975). Of the fourteen drag queen models in Warhol's project, Ross was the most photographed and the most painted.[165]

Finally, there is Charles Ludlum's the Ridiculous Theatrical

Company, which from the late 1960s performed numerous productions
in which men played the roles of women. Darling appeared fleetingly.
Ludlum claimed to give parts to 'drag queens off the street . . . if they
had outré wardrobe'. [166] Montez was a favourite: 'Montez captures the
ineffable essence of femininity', Ludlam observed. 'I never tire of writing roles for Mario Montez. He has dignity.'[167] It was the crossing back
and forth from masculinity to femininity that fascinated Ludlam, the
ability to play 'both sexes', and he explained that in one such role he
allowed his chest hair to show so that the audience would be clear that
'I was a man playing the role.'[168] It was drag rather than transsexuality
that appealed to the leader of this queer theatre company. Indeed,
Ludlam's *Bluebeard* (1970), with Bluebeard as a sex modifier, which
featured Montez as Lamia, the Leopard Woman, was, as its playwright
put it, a critique of 'altering people's bodies to match their sexuality –
"A man can't make love to a man; therefore we'll change one of you
into a woman."'[169]

Blurring Boundaries

There is an interesting 1990s interview in a collection called *Bodies of
Evidence: The Practice of Queer Oral History* (2012) in which Teresita la
Campesina, a Latina/Latino transgender performer, resists categorization even though her interviewer claims that the drag queens of the
1950s and 1960s 'denoted what later became transsexual and transgender'.[170] Teresita had worked behind a bar as a woman, 'with falsies,
looking *gorgeous*, like a sunset . . . working like a woman with all the
machos, sucking dick[,] fooling the men'.[171] However, she did not see
herself as a woman:

> Because I am man and woman. I receive; I don't penetrate . . . I know
> that even though I was born – I am never going to look like a woman
> . . . See, once that feeling is gone, you know what I'm saying? And if
> you like to masturbate, like *I* do, and I like to play with my cock daily,
> and I like my titties sucked, and I like to jack off, honey – I'm *not* gonna
> make a mistake . . . We are women of the *mind* . . . though you are a
> man and you got everything of a man . . . you don't have to cut off, and
> cut off the feeling . . . I am still the man and the woman.[172]

Teresita had considered surgery but had decided against it because of its perceived repercussions for sexual pleasure: 'When they told me that I was gonna be like a brand new Cadillac – *and no motor in it?* I'd rather keep this old jalopy[,] I still have a lot of sparks.'[173]

We have seen a blurring of boundaries from the early days of transsexuality. STAR, Street Transvestite Action Revolutionaries, an early trans activist group in New York in 1970, reflected this fuzziness of categories.[174] When Allen Young interviewed Marsha P. Johnson, a black sex worker and one of the group's leaders, in an essay published in a principal text of gay liberation, they shifted back and forth from referring to 'girlies', 'transvestites', 'gay people', 'gay transvestites', 'transsexuals', 'pre-operative' transsexuals, and 'women'. Johnson made it clear that STAR was part of the gay liberation movement – hence the interview's presence in the text – and that she was 'gay', 'proud', and a 'transvestite'. She lived her life in 'drag', survived through sex work in Times Square as a trans woman, and contemplated a future 'sex change'; but at the time of the interview saw herself and STAR under the framework of gay liberation: 'Our main goal is to see gay people liberated.'[175]

Benjamin reported that by 1966 only around 30 per cent of his 242 male-to-female transsexual patients had actually had surgery; presumably, then, most cross-dressed and took the hormones that he provided.[176] They would have been the 'large group' that his colleague Charles Ihlenfeld referred to as electing to stay in 'their anatomic role while taking hormones'.[177] Benjamin had also treated around 100 transvestites with oestrogen: not the heterosexual, married men who sometimes dressed in female attire, but those he termed 'bisexual' and requiring more than symbolic representation of their femaleness: 'They want to be aware of some physical changes, especially bust development, to support their feminine identification and ease their emotional distress.'[178]

Thomas Kando reported of Minneapolis in the late 1960s that, despite clinicians' separation of the transsexual from the transvestite, the reality was their 'similarity' rather than their difference. Indeed, he identified a 'subculture' of transvestism, transsexuality, and homosexuality based in the city's bars and stripclubs, with both transvestites and post-surgical transsexuals working as strippers and sex workers.[179]

As he put it, it was 'important to realize that the borderline between transvestites and transsexuals is not sharp'.[180]

Susan Stryker's interviews with transgender people who lived in San Francisco during the late 1960s and early 1970s, Driscoll's earlier-discussed milieu, were far from inexorably transsexual, indicating varied stories of partial transitions. Major – 'Just Major. I identify with both sexes' – who had been a 'street queen' in Chicago in the 1950s, said that 'There wasn't all this terminology, all this TS stuff . . . They were on hormones to soften their skin and their appearance, not because they wanted sex reassignment. They just wanted to better their look.'[181] Gina McQueen, who presented as a woman from the time she left home as a teenager, worked in a gay bar as a drag queen but also spent time in straight bars 'picking up straight men . . . Some I would lie to, some I would tell the truth to.'[182] She recalled not being able to grasp that some of 'those girls liked other girls . . . that warped me'.[183] She flirted with sex reassignment but ended up having 'a lot of plastic surgery' – face, nose, breasts, hips, legs, 'ass', 'just not genital surgery':

> I've been on hormones for years . . . I had decided that I wasn't going to be a woman, so I might as well be a sex goddess . . . At the same time, I decided that if I wasn't going to be a woman and be a sex goddess instead, that I was going to be a sex worker . . . Thought: hate men, love cock. Young men paying a hundred dollars a throw. Ha! Sounds good to me. Did it. Enjoyed it.[184]

The breasts, she claimed, were 'for money': 'Tits are always a money-maker, you know? That's what men like. And I've got them because of that.'[185] She was clearly comfortable in the postmodern context of the time of her interview (1997) when she was aware that 'I don't have to fit a stereotype . . . I don't have to get it cut off to get a pussy to fit into straight society which I'm never going to fit anyway.'[186]

Aleshia Brevard said that in the drag performance scene she was surrounded by those who were transgender but who 'just didn't want to go through surgery'.[187] She had worked in the famous San Francisco drag club Finocchio's in the late 1950s and early 1960s. She was then Buddy Crenshaw performing as Lee Shaw. Though she moved in 'the gay subculture' and knew that she was not heterosexual, Crenshaw also felt different from 'the young gay drag queens with whom I worked. I wasn't lampooning women; I was trying to find myself. Drag . . . was the closest I'd ever come to feeling that I belonged'.[188] 'I was just "femme"', she responded to an interviewer who pressed her about her early identity: 'I was part of that gay society, but at the same time, I wasn't.'[189] It was the transitionary world referred to earlier. 'Most drag queens are faggots earnin' a living', a fellow drag performer told Brevard, 'but a few are real women in male bodies and sequin dresses. You've gotta decide which one you are.'[190] Brevard, who self-castrated because she could not afford surgery, thought in retrospect that she might never have gone through the vaginoplasty that she eventually was able to pay for if it had not been for the pressure from her then boy-friend and being caught up in a moment when 'society sort of pushed us along'.[191] She resisted any suggestion that it had been a logical process. She implied that she felt good being feminine and enhanced that per-sona, piecemeal, starting with the female impersonation, and then the hormones: 'See, those of you that have some sort of logical tracking in you[r] mind – huh-uh, I'm not one of those.'[192]

Doctors at the New York University School of Medicine monitored nearly a hundred 'non-operated', male-to-female transsexuals who were on public assistance and who visited the city's welfare centres for psychiatric consultation in the late 1960s and early 1970s.[193] They were, the doctors reported, but the tip of a much wider population of New York trans women, white as well as African-American and Puerto Rican (the monitored group came mainly from the latter two populations). Lloyd Siegel and Arthur Zitrin, the authors of the study, claimed a large subculture, 'a kind of transsexual underground', much like the just-noted Minneapolis one, that was self-sustaining in the absence of any surgical solution for most of its members. They met at sympathetic treatment centres and drag clubs (where 'transsexual–transvestitic fan-tasies' were created) and were often roommates or moved in small groups who lived near each other and formed, as the authors put it, 'a kind of network'.[194] They swapped beauty tips and information about hormonal treatment and depilation. They interacted in a manner that constantly reinforced a mutual femininity: 'This unending interest in creating a feminine environment, style and appearance became, for a large number, their raison d'être.'[195] The interest in this group is its transsexuality without surgery: 'The feeling of really being a woman did not come to depend on anatomical reality or surgical alteration . . . For many, illusion created by group validation served as a substitute for surgery.'[196] Siegel and Zitrin called them transsexual in 1978. They would later be termed transgender.

Suzy Cooke, a trans woman, provided a visual chronicle of life in Hollywood ('Trannywood') in Los Angeles in the early 1970s – what she termed 'a golden age of relative ease of transitioning and in day-to-day life'; 'At that time the drag and trans scene was all mixed together. People lived full-time as women and performed at drag bars for money.' She thought that, although there was an awareness of transsexuality, it was not predicated on surgery – 'not so much in terms of having surgery, but just in terms of having different priorities'. Cooke claimed that it was easy to get hormones and breast implants.[197] As for blurring the boundaries, one of her pictures is of two trans women kissing, Cooke and her 'sister lover'.[198] Cooke later told Stryker that she 'slept with a lot of tranny girls'.[199]

The sexual mix of this era is clearly reflected in the personal advertisements of the late 1970s, notable for their varied combinations. 'Bi-sexual . . . Want to date generic females and/or FTM, TVs or men . . . Enjoy the fem role', wrote one correspondent. Vikki O., aged 28 'and just out of the closet', was interested in 'dating men, women and other TVs while crossdressed. Into all forms of erotic bondage.' Denise H. was 'Divorced, bi-sexual with other TV girls . . . Want to meet other TVs for dates.' Jodie K. was a 'Happily transgendering writer. Bi-sexual . . . Enjoy being fem with a man. Can be active or passive with a TV or TS. Would absolutely love to date a pre- or post-operative FTM.' Sunny K. was a self-proclaimed 'Statuesque 30-year-old feminist lesbian. MTF seeking women or TS also feminine for loving warm lesbian relationship.' Meanwhile Pat, 'Hetero, closeted', wanted a 'young TV/TS or tall FTM or attractive eccentric female for lasting relationship'.[200]

Those were all from the San Francisco area, publicized by an agency for MTF and FTM clients. However, the publication *Drag: A Magazine About the Transvestite* cast the net wider in its personals. Hence, the DC Area male who wanted to meet a 'feminine she-male, F.I. [female impersonator], TV, drag queen, pre- or post-operative TS for fun and friendship. Gentle is my name, pleasure is my game.' In

North Carolina, there was an 'Effeminate petite DRAG QUEEN with shapely legs, pretty feet, hot tongue. Expertise of foreign arts. B/D, lesbianism.'

In Philadelphia, 'Two foxy and sexy queens' advertised themselves alongside their pictures, two big-haired TVs in suspenders: 'will date males, females, TVs and all. We are discreet and real. Also dig bizarre interests!' In New York, a 'Tall, slim, attractive bi-TV' sought meet-ings 'with other bi-TVs. Have special interest in films of wrestling lesbians.' Finally, in Detroit, a 'Bisexual handsome male would love to meet shemale, TS, F.I., drag queen or TV for fun and friendship. And female who would like to dress me in lingerie.' And these examples come from only one issue in 1977.[201] The point is that the bodily variety observed earlier was accompanied by equally diverse desires.

The source material is rich, then, and it is curious that the history of transgender has not made more of it. Just as they would again in the 1990s (see Chapter 5), the photographers Nan Goldin and Amos Badertscher (independently) conveyed the gender and sexual indeter-minacy that I have been discussing. Goldin's immersion in the lives of the drag queens of the Boston bar The Other Side resulted in a portfolio of intimate portraits of male femininity. 'I was eighteen and felt like I was a queen too', she recalled later; 'Completely devoted to my friends, they became my whole world. Part of my worship of them involved photographing them. I wanted to pay homage, to show them how beautiful they were. I never saw them as men dressing as women, but as something entirely different – a third gender that made more sense than the other two.'[202] The beguiling photograph 'David at Grove Street, Boston 1972', comes from that period. See Illustration 16.

Similarly, Amos Badertscher's photographic study of Baltimore street and bar life from the 1970s onwards includes the sexually androg-ynous and ambiguous.[203] Branka Bogdan and I have written about Badertscher's trans models elsewhere, but it is worth revisiting some of the images.[204] They presented transgender in its varied configurations. He knew Steven, an androgynous, waif-like individual, from that trans-sexual moment of the 1970s. See Illustration 17. This working-class, teenage hustler, who told Badertscher that she considered herself a woman, acquired breasts, and transformed herself into the far from fragile Marilyn Mansfield, performing first in topless-bars and then

16 *Nan Goldin, 'David at Grove Street, Boston 1972'.*

17   *Amos Badertscher, Steven, 1973.*

in trans pornography.[205] Then there was Sandy, who died in the late 1970s, not long after her photograph was taken. See Illustration 18. She was a 'very sweet person who frequented gay bars . . . liked tight dresses and wanted very large breasts. The silicone injections led to epidermal dysfunction and cancer.'[206] Then there was Todd, whose make-up and cross-dressing was part of his teen hustling. One picture was of him in 1975, at the age of 21, a 'young Catherine [*sic*] Hepburn'.[207] Another,

18   *Amos Badertscher, Sandy, 1976.*

reproduced here, is of him on a bed, in stockings and suspenders. See
Illustration 19. Frank's was the first photograph of a male in drag that
Badertscher took – and he classified it, in the context of the 'gay scene'
of the early 1970s, as experimentation. 'Neither Frank nor I knew for

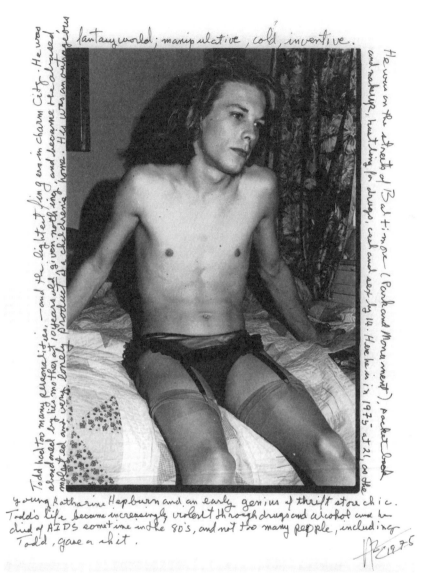

19 *Amos Badertscher, Todd, 1975.*

sure what we were doing here – we were both exploring and enjoying it.'[208] See Illustration 20. The only fully transitioning transsexual was Miss Natalie, whose 'dream was to become a woman'. Badertscher noted in the text that, once she had transitioned, Natalie wanted

20 *Amos Badertscher, Frank, 1974.*

no reminder of her maleness – which, of course, the photographs were.[209]

The Baltimore images work variously. They are all posed: it would be inappropriate to describe them as documentary. Sometimes, the sheer whiteness of the background, the lighting, forces focus on the black-and-white photography of body and accompanying text. Yet, in other instances, the mise-en-scène, the intimacy of a bedroom, draws in the viewer. To quote Tyler Curtain, 'Badertscher's art takes the feel, the excitement, of a trick and transfers it to the surface of the photo.'[210] However, we are more concerned with the gender ambiguities portrayed than the invocation of any sexual encounter: Steven's blouse and androgyny; Sandy's long lashes and breasts; Todd's suspenders, his foundation, plucked eyebrows, and lipstick; Frank's mascara, earrings, and nail polish.

Conclusion

Perhaps John Money best captured the milieu that we have been discussing. 'I do not really think that we can make a diagnosis in the differential sense that one can in many other aspects of medicine: this is a transsexual; this is a transvestite; this is a very effeminate homosexual, or a very virilistic lesbian', he wrote in 1978, when he moderated a discussion for the *Archives of Sexual Behavior*; 'I believe that we are dealing with a spectrum. There are transition zones in which it is very difficult to get the absolute clear image. Many patients occupy the transition zones of the spectrum.'[211] In New York, too, the files of the psychiatrist Ethel Spector Person indicate that she and her colleagues had difficulty distinguishing between transvestism and transsexuality. 'I would therefore feel that this patient presents more of a picture of a transvestite than that of a transsexual, although the lines of demarcation can never be clearly drawn', was the verdict on one such patient.[212] So the clinicians most associated with these taxonomies had doubts about their categories.

Stoller's *The Transsexual Experiment* (1975) begins by distinguishing between transvestism, effeminate homosexuality (he was talking about those designated male at birth), and transsexuality. Yet, despite this rather clear-cut categorization, Stoller had another group, 'The Mixed Group'. Some were fetishistic cross-dressers who *did* want bodily modification; 'Others are homosexual prostitutes who can pass as pretty women as well as transsexuals can but who enjoy their penises and do not wish to be females.'[213] In an earlier publication, he wrote of transvestites with transsexual tendencies, of men (*sic*) happy to develop their breasts but who 'unquestioningly draw the line about modifying their penises', of those 'who want to be publicly known as homosexual men who have been transformed into the appearance of women but neither they or their partners feel that they *are* women, but rather simulated women'.[214] They would have included people like the trumpet player whom a colleague at Cedars-Sinai Medical Center, Los Angeles, described in 1973. They had taken female hormones but seemed to hover between transvestism and transsexuality. 'He [*sic*] is aware of the ambiguous overlapping of his [*sic*] wishes to be a woman, to be like a woman, and to be with a woman – but "I'm confused and I love it".' The report concluded by raising the possibility of an 'area between transvestism and transexualism [*sic*] in which there is a strong wish to feel like a woman all of the time and without a willingness to undergo surgical removal of the penis'.[215]

Then there were those whom Stoller described as 'a complex mix of fetishistic, homosexual, heterosexual, and transsexual tendencies, all of which may be present at one time while at another only one style will predominate'.[216] His mixed group included the family 'man', a pharmacist, who intermittently self-administered oestrogen for 'his' breast development and who wanted 'his' genitals removed but was ambivalent about having a constructed vagina, was unconcerned about 'his' facial hair, and did not want to pass as a woman.[217] Alternatively, it included a 'man' who had removed all 'his' facial hair, who dressed and lived as a woman, and who took oestrogen to enhance 'his' female-ness, but who had no desire to remove 'his' penis and did not want 'sex change' surgery: 'He [*sic*] does not consider his body to be female; he [*sic*] does not believe himself [*sic*] to be a genetic trick of fate . . . neither in his [*sic*] conscious fantasies nor in his [*sic*] dreams does he [*sic*] sense himself [*sic*] as a female psyche trapped in a male body.'[218] We are not talking here of a few individuals. There was a sufficient number of them to warrant description as a group, and Stoller said that his research team in Los Angeles saw them in 'much greater numbers than we see transsexuals'.[219] He was aware that they threatened the neatness of his classifications. He conceded that some 'partially fit criteria of both transsexualism and transvestism'.[220] Yet he still maintained the usefulness of his diagnostic categories, claiming that they were in fact confirmed by such breaches.[221] The 21st-century reader may be less convinced.

Ethel Spector Person's files contain a continual slippage between homosexuality, transvestism, and transsexuality. The subject of one case history maintained that 'There is now a feeling by some T.V.'s that would like relations with a woman or very attractive T.V.'s as woman to woman and call themselves male lesbians. I must confess that if I were physically capable I could also go that route.'[222]

There is an argument to be made, then, that the sexual and gender flexibility normally associated with the transgender turn after the 1990s – and the subject of a later chapter – was anticipated in the trans-sexual moment. In fact, Cockettes-like 'genderfuck' was part of a wider counter-culture in the early 1970s, represented in David Greene's pho-tographs of the street theatre of Bono, Harmodius, Christopher Lonc, and Tanye, hirsute men in dresses, who saw their gender mashing as a critique of straight society.[223] As Lonc wrote, 'I want to try and show how not-normal I can be. I want to ridicule and destroy the whole cos-mology of restrictive sex roles and sexual identification.'[224] Genderfuck, in the form of 'radical drag', was also a technique of 'extended guerrilla theater' employed by radical gay protestors challenging gender role ste-reotypes.[225] Betty Luther Hillman has untangled the tensions between nascent trans activism and more traditional homosexual politics in the San Francisco gay liberation movement in the late 1960s and early 1970s.[226] Yet the languages associated with transgender were part of radical discourse. 'When a Street Queen walks down the street', wrote a leader of the Berkeley Gay Liberation Front, mentally skipping the transsexual moment, 'she's a one-person rebellion, her being an affront and attack on the straight world around her . . . She refuses to be either a man or a woman, and that is the most revolutionary act a homosexual can engage in.'[227] Trans visual imagery made an appearance, too. Jackie Curtis was on the 1970 front cover of *Gay Power*, a New York gay newspaper.[228]

The memoirs of the rock singer Jayne County likewise suggest some continuity between the periods of transsex and transgender. The then Wayne County's experience as an effeminate gay man in Atlanta – 'a Screaming Queen' – merged with her later New York drag queen days, living with the Warhol celebrities Curtis and Woodlawn in the late 1960s, pondering about her transsexuality in the 1970s, taking hormones, and then later, as Jayne County, her adjustment to transgender as a woman with a penis: 'I'm happy in between the sexes.'[229] She recalled a time when 'being different' was more important than being classified as 'gay, lesbian, transsexual, bisexual, transvestite or anything'.[230] County's cross-dressing performances were always confronting in a Cockette-like way, gender antagonizing rather than gender blending. Though still using the term 'transsexual' when she published her memoirs in 1995, County was of the transgender turn before there was a transgender turn.[231]

4

Backlash

Introduction

'This passing fad for what is miscalled "transsexualism" has led to the most tragic betrayal of human expectation in which medicine and modern endocrinology and surgery have ever engaged', the prominent New York psychoanalyst Lawrence S. Kubie wrote in 1974. He was pulling no punches. 'In the name of gender transmutation they have led people to believe that alchemy was possible, thus fostering in individuals and in our whole culture conscious and unconscious neurotogenic fantasies whose only possible outcome is an intensification of the neurotic fantasies which underlie their expectation and ultimate psychosis.'[1]

We have discussed the rapid rise and acceptance of the term 'transsexuality' at the time of Harry Benjamin's *The Transsexual Phenomenon* (1966), Robert Stoller's *Sex and Gender* (1968), and Richard Green and John Money's edited collection *Transsexualism and Sex Reassignment* (1969).[2] But we also hinted at a lesser-known and darker history of rejection and questioning, reflected, rather outrageously, in Kubie's outburst. Joanne Meyerowitz, transsexuality's skilful historian, is aware of what she terms 'years of unresolved tensions', yet devotes relatively little space to such strains.[3] Susan Stryker, transgender's best-known chronicler, has not dealt with the issue at all.[4] The result is not exactly a case of hidden history, but rather inattention to an important period of critique of what Benjamin called the 'Transsexual Phenomenon'. This neglected story both rethinks transsexuality's brief history and sheds rather negative light on US psychiatric, surgical, and therapeutic practices in the 1960s and 1970s.

Early Critiques

Transsexuality's counter-history can be traced back to the very beginnings of the emergence of the term and its earliest treatments. The psychoanalysts seemed always to be wary. A New York therapist wrote in the letters section of the *American Journal of Psychiatry* that, with many transsexuals, physicians were unwittingly treating fantasies rather than diseases: 'Those patients seen by me who wanted to change their genital status were all borderline psychotics who also wanted other parts of their body altered.'[5] *The Journal of Nervous and Mental Disease* featured a piece co-authored by the aforementioned Kubie that was highly critical of the ready acceptance of the term 'transsexual' and advocated 'eschewing' the word altogether. The authors thought that 'gender transmutation' better described the processes involved, whether endocrinological or surgical.[6] The simplicity of the term 'transsexual', they argued, masked the complexity of the problem and implied 'that unsolved problems have been solved'.[7]

The article listed the issues glossed over through a simple clinical diagnosis of transsexualism. There were men who wanted to

> achieve the *appearance* of being women, but wish to think of themselves and be known as men who 'simulate women'. Such men often slant their descriptions because they soon become aware that in most medical centers, in the United States at least, they must present themselves as textbook examples of 'transsexuals', if they are to persuade any team of physicians to change them.[8]

Hence the transvestite became the transsexual. Also, the simplicity of the diagnostic model – what the authors termed the 'premature solidification of the concept of "trans-sexualism"' – forced conscious or unconscious conformity.[9] Personal histories were reformulated. Patients were tutored by friends, or their histories formulated by their own, targeted research.

> Even with the limited information currently available it is clear that *not all* patients who have undergone surgical changes were unalterably convinced of their membership in the opposite sex. Moreover, abundant clinical and some empirical data show that retrospective self-justification can play a role in distorting the developmental histories given by individuals who petition for sex reassignment.[10]

Kubie and his co-author, James B. Mackie, concluded that the concept of transsexualism provided the unique combination of false diagnostic and conceptual clarity with 'emotionally charged' and 'dramatic' medical intervention.[11]

In a short article, originally written for the *American Journal of Psychiatry* in 1973, Stoller also outlined 'uneasiness' about what he described as the 'carnival atmosphere that prevails in the management of male transsexualism'.[12] The medical climate had changed rapidly in a matter of years from a situation where opinion was divided over diagnosis and surgery to one of widespread acceptance based almost solely on the patient's mere request for 'sex transformation' – though the existence of his piece indicated another, less endorsing shift, the one that we are discussing here.[13] Like Kubie and Mackie, he believed that diagnosis was complicated by the familiarity of the supposed transsexuals with the medical literature. It was hard for therapists to determine 'how firm is the patient's conviction he really is a woman trapped in a male body', when those treated 'are in complete command of the literature and know the answers before the questions are asked'.[14] He thought that people with slender links to transsexuality and with ill-defined claims of femininity were being treated surgically and irrevocably as transsexual. 'We all know of surgeons willing to operate as long as the price is right; they seem scarcely concerned even when inexperienced.'[15] Furthermore, there was little or no concern with follow-up research. He concluded: 'Since 1953, when "sex change" procedures were first publicized, an unknown number of males has received hormonal and surgical treatment on request. That we have no notion how many have been treated, when the procedures are experimental and potentially dangerous, is astonishing. That we do not know, almost 20 years later, how the patients fared, is scandalous.'[16]

These professionals recognized patient agency, albeit negatively – an issue that recurs in this book. But there was early patient criticism of doctors as well. The trans woman Canary Conn outlined a scenario of humiliation at the hands of Los Angeles doctors in the early 1970s, after initial surgery in Mexico. She was displayed to interns without even an introduction, let alone consent: '"Real butcher job", one intern said . . . They all turned to him and laughed'. '"Now this", Dr. Wells said, grabbing my penis, "this will be used in the ultimate construction of the vaginal canal".'[17] Conn fought back the tears and the anger because of the possibility of treatment. But she decided after the protracted experience of counselling and prevarication that she was being used as a subject in a study of transsexuality, and that no surgery had ever been intended. 'I decided to stop dealing with the world of the American medical man.'[18] It was around this time that the trans woman and artist Erica Rutherford consulted a Pittsburgh practitioner who told her, 'Of course, someone of your physique would not make a very presentable woman.' 'I was struck dumb by his remark', Rutherford recalled years later, and it 'plunged me back into the world of shame and frustration where I had lived for so long'.[19]

Jane Fry's autobiography implies a discontent with medical expertise that slipped into contempt.

By this time I was getting very down on doctors. I am still very turned off by them. For one of the most educated professions they have some of the biggest assholes I have ever seen in my whole life. I went to see Dr. Moore. I sat down and went on with my normal spiel. By this time I had like a sales pitch. He said, 'So you're a transsexual, huh? Well, that's good. I was just reading some articles on that.' It just freaked me out. I didn't know what to say. He said that he was just reading Dr. Benjamin's book . . . At least he had read something and knew a little bit about what he was doing . . . Actually, he didn't know that much about hormones and their effects. It is very tricky . . . I knew more about hormone and therapy treatment than he did. I kept an eye out that he didn't give me too much or too little.[20]

This trans woman claimed to know more than her therapists and was cynical in her interactions.[21] As Fry's collaborator and editor stressed, her version was profoundly at odds with the perspective of the professionals. She was acutely aware of the centrality of the doctor in the lives of transsexuals – more important sometimes than the closest of kin. They prescribed the hormones and recommended surgery: 'One thing that I did learn in meeting all the doctors is that you have to give a little – pretend a little. Any one of them can kill you physically or emotionally.'[22] At the same time, she was scathing about psychiatric explanations for transsexualism: 'Every doctor you see gives you a different explanation, and you just come to the point of knowing that they just don't know [what] the hell they are talking about.'[23] One therapist told Fry that her desire to be Jane was because her father was so violent: 'in rejecting him I rejected masculinity and violence, so I had to be a female'; 'I think that's bullshit.'[24] In and out of psychiatric care after leaving the Navy, and suicidal, Fry submitted to therapy as a means to an end (surgery), with little conviction as to its efficacy.: 'I don't think my transsexualism is the direct cause of my emotional problems, but I have to let psychiatrists keep saying it or else they won't treat me. I have to get back on the road to getting my operation, so I have to see one.'[25]

These snippets of early patient and therapist comment are consistent in their critical themes, locating problems of what might be broadly categorized as attitude to patients, surgery, therapy, definition/diagnosis, and assessment. Let us examine these in more detail.

Attitude to Patients

One clear problem was doctor–patient interaction. A mid-1960s survey of the opinions of surgeons, urologists, gynecologists, general medical practitioners, and psychiatrists – the very people whom a transgender person might approach for help – found a general conservatism when it came to sex reassignment. Even when the most responsible of treatment regimes was proposed in a hypothetical scenario involving prolonged psychiatric consultation as a preliminary to bodily remodelling, only 37 per cent of surgeons, 45 per cent of psychiatrists, and 41 per cent of GPs would approve a request for sex transformation. The majority of practitioners were not exactly sympathetic.[26]

Some experts were simply hostile to the whole notion of transsexuality. The psychiatrist Charles Socarides thought that transsexuals
were delusional, and included homosexuals in denial who thought
that changing sex would render their same-sex desires acceptable as
heterosexuality: 'In this author's opinion . . . surgical intervention con-
stitutes a sanctioning of the transsexual's pathological view of reality
and cannot resolve the underlying conflict.'[27] The sociologist Edward
Sagarin (author, as Donald Webster Corey, of the pioneering 1951
pro-gay work *The Homosexual in America*) was similarly sceptical about
transsexualism, writing in 1969, in a book on 'deviants', that male-to-
female transsexuals suffered from 'doubly unacceptable' self-imagery
– that of being both homosexual and effeminate.[28] The obvious solution
for such 'victims', Sagarin posited, was for them to convince themselves
that they were really women not men: 'Thus, having sex with a man is
not an abnormal act for the transsexual because he is, in his self-view,
a woman.'[29] In reality, Sagarin wrote, 'normalcy' in any relationship
between transsexual and a male was 'impossible . . . the partner is and
must be a homosexual, and . . . even with conversion surgery, a transsexual can at most become a castrated male with an artificial vagina'.[30]
He observed of Fry's autobiography that it was 'a sad and dreary story
of a youth with strong effeminate traits, learning about transsexualism
and then deciding that that is what he was all along'.[31] One had only to
read Benjamin's case studies, 'to note how disturbed are the patients'.
Transsexuals were 'deviates' to be discussed alongside necrophiliacs.[32]

Socarides was notoriously anti-homosexual, and Sagarin/Corey was
conflicted about his sexuality. Yet even those engaged in pioneering
transsexual research and treatment could not avoid a battle-weary
negativity in their assessments of their patients. Hence, the notorious
case of Agnes, one of Stoller's patients, written up in a classic study by
the sociologist Harold Garfinkel in 1967.[33] Agnes, who first presented
in 1958 as a 19-year-old woman, a typist, was treated as an intersexed
patient. From the outset, the team from the Departments of Psychiatry,
Urology, and Endocrinology in the Medical Center of the University of
California–Los Angeles, accepted her as female, 'convincingly female':

She was tall, slim, with a very female shape. Her measurements were
38-25-38. She had long, fine dark-blonde hair, a young face with

pretty features, a peaches-and-cream complexion, no facial hair, subtly plucked eyebrows, and no makeup except for lipstick. At the time of her first appearance she was dressed in a tight sweater which marked off her thin shoulders, ample breasts, and narrow waist.[34]

But along with her large breasts (and exaggerated 1950s femininity), she had the 'normal external genitalia of a male'. She was a 'female with a penis'.[35] They carried out the kind of tests used for the intersexed – which revealed no internal female organs but 'moderately high estrogenic (female hormone) activity' – and, in retrospect, relying rather too much on the patient's testimony, concluded that this was an individual born as one sex but developing the sexual characteristic of the other at puberty, when the testes started to produce oestrogen, 'a unique type of a most rare disorder'.[36] The team wrote a learned paper, 'Pubertal Feminization in a Genetic Male'.[37] They arranged for Agnes (as intersexed rather than transvestite, homosexual, or transsexual) to have the surgery in 1959 that removed her penis and testicles, and used them to construct her vagina and labia. (Agnes had a boyfriend, Bill, who had been eager for her to gain a vagina so that they could have intercourse.) Then, several years later, in 1966, Agnes admitted to Stoller that she had been taking oestrogen from the age of 12 (stolen from her mother) and had lied to her doctors when they had specifically considered that possibility. The psychiatrist wrote, 'My chagrin at learning this was matched by my amusement that she could have pulled off this coup with such skill.'[38]

We would now see Agnes as an early example of patient agency. To quote Kristen Schilt, this was 'an account of a young woman who successfully navigated a medical system designed to "weed out" people like herself'.[39] But the point about this case, at the time, was that it reinforced notions of patient duplicity. Even before all had been revealed, Garfinkel's long account of their interaction indicated a certain amount of perceived dishonesty on the part of Agnes, 'how practiced and effective Agnes was in dissembling . . . she was a highly accomplished liar'.[40] Of course, Garfinkel was oblivious to the bigger lie and was treating her deception as an integral part of her passing in order to attain the femininity that she claimed was natural: 'In contrast to homosexuals and transvestites, it was Agnes' conviction that she was naturally, origi-

nally, really, after all female.'[41] But the parade of deceits outlined by the
sociologist were those that became the alleged tricks of transsexualism.

Because Garfinkel was so alert to what he termed 'management
devices', he was able to list off those employed by Agnes: 'shrewdness,
deliberateness, skill, learning, rehearsal, reflectiveness, test, review,
feedback'.[42] She was adept in patient–therapist encounters:

> When I read over the transcripts, and listened again to the taped
> interviews while preparing this paper, I was appalled by the number of
> occasions on which I was unable to decide whether Agnes was answer-
> ing my questions or whether she had learned from my questions, and
> more importantly from more subtle cues both prior to and after the
> questions, what answers would do.[43]

She presented a 'remarkably idealized biography', in which her femininity was exaggerated and her masculinity suppressed.[44] She refashioned
her personal history, 'reading and rereading the past for evidences to
bolster and unify her present worth and aspirations . . . she was engaged
in historicizing practices that were skilled, unrelieved, and biased'.[45]
She also knew when to remain silent and how to deflect. Garfinkel said
that, after seventy hours of communication with various therapists, and
despite specific questioning on these issues, there were areas – mainly
about Agnes's sexual interaction, including what she did with her penis
– 'in which we obtained nothing'.[46]

So, distrust of patients was there almost from the start. Witness
Elmer Belt's discouraging appraisal in 1969: 'I found that these patients
in general were so unsuited to handle the problems of life itself that
changing their sex organs was not a satisfactory solution of their trou-
bles with society.'[47] They were rarely satisfied with the results of their
treatment. They craved publicity of the kind that was anathema to
their doctors and medical centres. Many were oblivious to the codes of
public behaviour. Belt claimed that his transsexual patients so troubled
his non-transsexual clients that he had to establish separate premises for
their consultation. And – in what we have already noted would become
a familiar refrain – their reported patient histories were unreliable: 'In
the case of the transsexual . . . who has become accustomed to live a life
of deception the history is apt to be a figment of lies. Virtually every

transsexual does not tell the truth in answer to the doctor's questions.'[48] Stoller could be equally scathing. In one short published piece, 'The Psychopath Quality in Male Transsexuals', he fulminated over their unreliability, their lying, and the superficiality of their relationships. He said that his team could almost diagnose a patient's transsexuality unsighted if they failed to turn up on time for the first appointment.[49]

So, even those who might be assumed to be the most empathetic, given their expertise and case-loads, could still betray a sense of disdain. Indicating his attitudes both to women and to male-to-female transsexuals, John Money and one of his (numerous) co-authors claimed that not only were 'male transsexuals [sic] . . . devious, demanding and manipulative' but they were possibly also incapable of love.

> Though there is no definite proof at the present time, it is quite likely that one of the characteristics of the transsexual condition in males [sic] is impairment of the neuropsychologic mechanism that mediates the experience of falling in love. If such be the case, then the full-blown experience of being in love is replaced, in the male transsexual [sic], by what, at the outset at least, is a more perfunctory, instrumental and opportunistic relationship. It may well be that the transsexual male [sic], when first reassigned as a female, is erotically 'turned on' more by the subjective imagery of having a functional role as a coital female than by the erotic stimulus-image of the partner or prospective partner.[50]

They wrote of the inadequacies of the surgically constructed vagina, the sexually anaesthetizing effects of encountering a penile stump, and, most disconcertingly, drew comparison with a study of a male turkey attempting to copulate with a head that had been separated from its mate's body. 'Like incomplete bird models, the human transsexual male [sic], though an incomplete and impersonating female, obviously projects at least the minimum number of feminine cues to attract the erotic attention of a normal male.'[51] The article was about the male partners of these 'incomplete' women.

L. M. Lothstein, the author of a pioneering study of female-to-male transsexuality that drew on data on over fifty such patients at Cleveland's Case Western Reserve Gender Identity Clinic in the 1970s and early 1980s, diagnosed female-to-male transsexuality as

'a profond psychological disorder'.[52] 'Most female transsexuals [*sic*]
... have serious personality disorders and while not psychotic, they
have subtle thought disorders which affect their sense of reality and
their ability to relate to others. In addition to their personality dis-
turbances many female transsexuals [*sic*] exhibit a wide range of other
psychiatric symptoms: including depression, anxiety, panic attacks,
and severe psychosomatic complaints.'[53] Lothstein saw the solution
to transsexuality as lying with psychotherapy rather than surgery, and
when advocating the latter did so reluctantly in service of the former.
Some trans men, he thought, were only amenable to psychotherapy
after surgery had 'disrupted their rigid defensive structure'![54] His key
words for trans men were 'failure', 'self pathology', 'impairment', 'dis-
turbance', 'defect', and 'developmental arrest'.[55] Such people, he wrote
in his 24-point strategic guide to therapy, had 'failed to develop a
core female identity ... her [*sic*] ultimate defect is a psychological one
related to her [*sic*] lack of a nuclear female self and a cohesive gender-
self system'.[56]

Moral judgement continually crept into diagnostic classification.
The Gender Identity Clinic of The Johns Hopkins Hospital divided its
1960s patients into two – morally classified – groups: the first, clearly
less-favoured, flamboyant, effeminate, homosexually oriented, 'antiso-
cial', petty criminal (a description hinting at the transvestite sex workers
and cross-dressers who sometimes transitioned to transsexuality); and
the more morally acceptable, better-behaved, more socially integrated,
second group that 'usually has tried, unsuccessfully, to make a heterosexual adjustment'. The practitioners argued that this classification
had ramifications for treatment: 'The patients in the second group
give more reliable histories and are much less manipulative, hysterical
and demanding than the first group.'[57] Whether, in practice, it was
always easy to separate the groups is another issue. Interestingly, James
Driscoll, who wrote a sociology of transsexual sex work, claimed in
1971 that two of his subjects were former patients of Money at Johns
Hopkins and had provided their specialist with very select histories: 'In
telling him their life story, they left out much and presented themselves
as a combination of Alice in Wonderland and Rebecca of Sunnybrook
Farm', while supporting themselves by sex work in Baltimore during
their treatment.[58]

Surgery

One of the great puzzles for non-trans people is why, knowing the trauma involved in reconstructive surgery (and transsexual patients were often very well-read), anyone would embark upon such a shattering experience. The earlier transsexual memoirs were graphic. Coccinelle, the famous French transsexual, recalled, 'I was in pain, terrible pain . . . I suffered a kind of martyrdom.'[59] Canary Conn said that she wanted 'to celebrate the occasion' but her body was convulsing in pain and she was screaming.[60] Nancy Hunt, who had her surgery at the University of Virginia in the mid-1970s, wrote both of shock at the carnage of the surgery and the 'unremitting pain'.[61] Female-to-male Mario Martino referred to the death of his new penis, 'black and foul-smelling', and recalled sitting in his bath cutting away dead tissue – 'Talk about castration complex!'[62] Trans woman Aleshia Brevard was 'in agony' after her surgery in the early 1960s; her new vagina, when she saw it for the first time, looked 'like something you'd hang in your smokehouse . . . after a hog killing'.[63] She was equally blunt in her summary of her surgeon's technique: 'In creating vaginal depth, Dr. Elmer Belt skinned the penis, took a skin graft from the bottom of each foot, and removed a six-inch square from the back of my left thigh.'[64] Patricia Morgan consulted the same surgeon and recalled almost identical experience of extreme pain. 'I was a woman at last. But at the moment, I was just a gob of aching flesh.'[65] Renée Richards referred to a 'bath of suffering':[66] 'It was as if someone was repeatedly poking a firebrand into my groin. Mixed with this was a tearing sensation; it was like someone was ripping at my organs with a pair of pliers.'[67] Although, overall, these were stories of successful transitions, they were not trouble-free journeys.

The academic papers also contained explicit illustrations of flesh-cutting and bodily separation and descriptions of the surgical aftermath. Garfinkel's study of Agnes contained a subtextual horror-story of vaginal closing, painful dilations, urethral contracture, cystitis, 'uncontrolled seepage of urine and feces', pelvic pain, and (understandably) 'sudden uncontrollable spells of crying' and depression.[68]

The state-of-the-art, early summaries of transsexual research of the late 1960s and early 1970s provided trans women with details of the process that reconstructed their genitalia, with a vaginal space cut and the penis turned inside out and then severed to make a 'functioning' vaginal cavity (strangely reminiscent of the sixteenth-century medical drawings of the vagina as an inverted penis), the urethra separated and repositioned, and the testes removed.[69] They were warned of the possibility of urethral stenosis (that is, narrowing), of the dangers of neo-vaginal closing, and the importance of the painful vaginal form or stent (balsa wood, lucite, or silastic) worn to prevent closure and held in place by an unglamorous girdle and a variety of plastic brackets, hooks, and elastic reinforcement – all pictured in black-and-white photographs.[70] 'Above all', wrote the Professor of Gynecology and Obstetrics at Johns Hopkins, transsexuals needed 'to understand that the reconstructed genitalia may not appear to them as perfect as they had envisioned. Many patients desire a perfect body form. They must clearly understand, prior to operation, that the postoperative appearance of their genitalia may fail to be quite up to their expectations.'[71] The illustrations in the published proceedings of a 1973 symposium tended to reinforce such warnings.[72]

The equivalent surgical summary relating to trans men referred dramatically to 'obliteration of the vagina, breast amputation, and hysterectomy'. But it also made it clear that those with – it should be pointed out, rather limited – clinical experience (six female-to-male patients in twelve years at Johns Hopkins) thought them less eager for the complete surgery demanded by trans women and more willing to agree to a 'gradual transition' and ready to 'accept imperfections'.[73] 'Selection of breast amputation as the primary procedure is in keeping with the patient's major quest and, in addition, seems less of a surgical assault than combined one-stage hysterectomy and beginning construction of a phallus.'[74] That 'surgical assault' is an appropriate description is reinforced by both text and illustration. 'Breast amputation' involved the grafted relocation of the nipples, and, in one case when the graft did not take, replacement of the areolae with sections of the labia minora. As for the construction of male genitalia, even the surgeons were sceptical:

> The patients fully understand that the phallus will serve little, if any, role in sexual activities; and it is hard to comprehend that they insist on undergoing the multiple hospitalizations and relatively extensive

operative procedures when they are aware of the severe limitation of this technique ... Creation of male external genitalia is not to be undertaken without deliberate planning and a substantial background of plastic surgical experience. One must be prepared fully to deal with the complications attendant upon loss of free grafts and pedicle flaps, and one must be conversant with the management of urinary problems in the form of fistulae, urinary tract infections, and incontinence. A considerable volume of experience is reported in the literature on the management of traumatic avulsions [tearing] of the penis and scrotum.[75]

The full sacrificial horror of constructing a non-functioning penis from an abdominal flap, and the resultant bodily scarring to achieve this early form of phalloplasty, can only be conveyed visually.[76] 'The bottom line with "bottom" surgery', an informed patient wrote more than thirty years later, 'is that no surgeon can give a transman the penis that he should have had at birth'.[77]

Therapy

Jane Fry's version of her case conference was unquestionably disheartening. One doctor summed up his opposition to transsexual surgery as 'If someone wanted to be an umbrella, would you try to make him an umbrella?' Others recommended aversion therapy.[78] That was her account, but her editor obtained Fry's medical files (in itself an indication of attitudes towards patients) and they showed that the professionals who dealt with her thought that they were treating someone who was psychologically disturbed:

> It is obvious that Jane has problems with sexual identity. It seemed to many of the staff that one of the core conflicts was around excessive dependency needs. She has been able to see that her creating crisis situations – taking overdoses – are often attempts on her part to have others express love and concern for her. In addition the role of 'poor Jane trapped in a man's body' is part of her repertoire. Underneath this, one might suggest that being a woman means being passive, being taken care of and not being active in satisfying one's need but getting that need satisfied indirectly.[79]

She was suppressing her masculine identity 'as a defense against castration anxiety'. The patient exhibited 'immature verbalizations around the topic of a new sexual identity . . . a pattern of character disorder of a very primitive nature'.[80] Her suicide attempts reflected the bad 'he' at war with – attempting to destroy – the worthless 'she'.[81] These various recorded diagnoses do little to inspire confidence in American psychiatry during the late 1960s and early 1970s.

The literature of transsexual treatment at that time makes for disturbing reading, with the psychiatrist's overreliance on the patient's outline of their early life, and the construction of shaky interpretations on these unreliable memoirs: 'at age 7 he plunged a knife into his father's prize cabbage . . . "I liked defiling the cabbage . . . I must have hated my father."'[82] There were blanket condemnations of parental roles, again based on the flimsiest of actual evidence: 'There was constant and excessive physical contact between the mother and child which the patient describes as, "I was like a monkey clinging to its mother."'[83] These quotes are from research from New York's Payne Whitney Psychiatric Clinic that presented case material open to pretty much any interpretation, but which was read as leading inexorably to the making of a 'sissy' destined to male-to-female transsexuality.[84]

Literature from the Gender Dysphoria Clinic at the Case Western Reserve University School of Medicine in the 1970s, which employed psychotherapy in the treatment of transsexuality, makes it difficult to understand how its programme could have been at all successful – though perhaps explains the large number of reported patients who left before the completion of the study.[85] Lothstein's discussion of the varieties of countertransference at work in the (multiple) gendered interactions between patient and therapist are interesting in their recognition of the complexities of such treatment, yet are notable more for the revealed attitudes towards the treated ('severe character pathology and bizarre self-presentation') and the perceived insecurities of the supposed experts ('disgust, jealousy, rage, terror and pleasure').[86] We read of trans women who were allegedly depressed when they were confronted by a '*real* woman' in the form of their female therapist, and of trans men who felt more masculine than their male doctors.[87] We are informed that the gender dysphoric patient is a 'guarded, secretive individual who uses massive denial, fails appointments, blocks, is superficial, narcissistic, employs primitive and morbid psychological defenses and bizarre self-presentation'.[88]

But even more unsettling is the perceived state of their therapists. Here we learn of the 'politically active gay female therapist' who may 'unconsciously' support the castration of the trans woman 'for non-clinical reasons', or, alternatively, the female therapist jealous because her male-to-female patient is a more attractive woman than she is.[89] A male therapist may become uncomfortable in the presence of a trans man who seems more of a man than he is, or may be condescending if 'she' fails to present the proper masculinity.[90] Lothstein thought that the 'construction of a phallus in the anatomical female stirs up the [male] therapist's own wishes and fantasies for a bigger and better penis', and, if the psychiatrist is a woman, that the patient's 'wish for a penis may stir up the female therapist's own pre-oedipal wishes for a penis and reactivate formerly resolved conflicts'.[91] These poor therapists must have required constant therapy.

We will see shortly that Lothstein's work raised questions about the fragility of some initial diagnoses. However, he (unintentionally) also demonstrated the potentially deleterious effects of over-analysis. When he wrote of the 'importance of intensive psychological testing and psychotherapy in the overall evaluation and treatment plan for the female transsexual [sic]', he really did mean intensive.[92] His cases were subjected to what he termed the 'full battery of psychological tests': the Minnesota Multiphasic Personality Inventory (MMPI), the Weschler Adult Intelligence Scale (WAIS), the Thematic Apperception Test (TAT), and the Rorschach.[93] Testing occurred both before and after sex-reassignment surgery – a year after surgery if that surgery was approved – and references to the patient's 'façade' and 'defensive structure' imply that the therapist assumed an eventual cracking of this surface to reveal deeper anxieties: 'she [sic] had erected a bland façade to control her inner excitement and confusion'.[94] It is little wonder, then, that the therapist's relentless testing eventually found this assumed disorder: 'In summary, the findings suggested a bi-modal clinical picture. On clinical interviewing and objective psychological testing Tina appeared "stable" and free of serious psychopathology. Intensive psychological testing, however, suggested that she [sic] had a subtle but non-intrusive thought disorder and the capacity to regress under stress.

The overall clinical picture suggested a mild borderline personality disorder.'[95] Little wonder, too, that this patient became 'depressed' and that a friend commented that they 'seemed more disturbed than ever'.[96]

Another patient, labelled 'Donna/Douglas', who took male hormones, bound his breasts, sported a goatee, and successfully presented as a man (known on the streets as 'Frenchy', 'a nickname which identified her [his] special talents for oral sex'), was considered to be troubled rather than transsexual and therefore not a suitable candidate for surgical reassignment.[97] Lothstein used this case as an example of someone who might have been carelessly misdiagnosed without the intensive evaluation that he provided. However, his formulation was influenced by Douglas's deteriorating performance after prolonged testing, as initially good results declined in later tests. 'Under stress Donna's [Douglas's] thinking became confused, illogical, and derailed, marked by tangentiality, symbolic meanings, and autistic logic.'[98] While the therapist interpreted this as evidence that 'most female transsexuals [*sic*], who appeared "healthier" on the initial clinical interview, might have severe character pathology', it is conceivable that the pressure of testing produced the detected psychosis.[99]

Then there was the team from the Department of Psychiatry and the Gender Identity Research Treatment Program, UCLA School of Medicine, Los Angeles, who treated 'pretranssexuals' in the late 1960s and early 1970s.[100] The notion of the 'pretranssexual' was based not only on the retrospective evidence of adult transsexuals who always interpreted what was seen as the excessive feminine behaviour of their own childhoods as an inexorable marker of their gender disjunction and eventual adult transsexuality, but also on the earlier work of Money, Richard Green, and others who used Konrad Lorenz's classic research on imprinting in ducklings to explain very early gender role learning in infants.[101] Green and Money had, in the late 1950s and early 1960s, claimed that early indulgence of effeminacy in boys could lead to transvestism and homosexuality.[102] (Note both the equation of effeminacy and homosexuality, and the absence of transsexuality, in this early work.)

A decade later, with Green now in the UCLA team, such children were seen as 'pretranssexuals' – hence, the 'very feminine young boys' who were in treatment from the age of 5 years old and younger, because of perceived 'feminine interests'.[103] It was reported of one 'transsexual' 8-year-old boy that his clinical history

> [p]aralleled the retrospective reports of adult transsexuals, including (1) feminine voice inflection and predominantly feminine content in speech, (2) verbal self-reference as 'sissy' and 'fag' and statements about his preference to be a girl, (3) feminine hand and arm gestures and 'swishy' gait, (4) an aversion to masculine play activities, (5) a strong preference for girl playmates and taking a feminine role in play and role-playing, and (6) improvised cross-dressing.[104]

George Rekers and O. Ivar Lovaas, who were prominent in such behavioural treatment (the above quote is from one of their studies), claimed that modification was most effective if the patient was under the age of 7. And they patently saw their goal as 'preventative treatment for extreme adult sexual deviations or transvestism, transsexualism, or some forms of homosexuality'.[105] (Homosexuality (as effeminacy) and transness were treated as synonymous in these studies.) Therapy in such cases included: 'Developing a close relationship between the male therapist and the boy, stopping parental encouragement of feminine behaviour, interrupting the excessively close relationship between mother and son, enhancing the role of father and son, and generally promoting the father's role within the family.' Doctors Green, Newman, and Stoller claimed success: 'Results indicate the capacity for gender role preference in the preadolescent male to undergo considerable modification toward masculinity.'[106] Carl, the previously mentioned 8-year-old, was dissuaded from modelling himself on the cross-dressing, early 1970s, very funny, African-American comedian Flip Wilson and encouraged to talk about masculine things ('e.g., firemen' and 'camping with the Boy Scouts') – did these therapists have no cultural awareness, no sense of humour?[107] A hapless research assistant became Carl's 'buddy' in the absence of the obligatory father figure: 'Interactions between Carl and the assistant included informal athletic sessions, tumbling lessions [sic], trips to the park, regular treats (e.g., sodas, ice cream), and occasional trips to the beach.' Apart from providing an exemplar of 'appropriate masculine behavior', the poor assistant also became 'a sounding board for Carl's many and bitter complaints about his past treatment by family members, school mates, and authority figures'.[108]

Karl Bryant (not the same Carl), one of Green's child study participants, who later wrote a Ph.D. on the history and context of the treatment that became known as GIDC (Gender Identity Disorder in Childhood), recalled a disapproval of feminine behaviour. Bryant quoted from a typical interaction – more a monologue – described in Green's book *Sexual Identity Conflict in Children and Adults* (1974): 'You know that sometimes little boys do wish that they had been born girls, and that's a little bit sad. One reason is that it's not possible for little boys to become girls, and it's sad because there are so many fun things that boys can do. They can go places and do things more on their own.' This, Bryant claimed, typified his interactions with Green.[109] Green asked the same 4-year-old whether his 'tweener' ever got 'kind of big and stiff' and whether that happened 'when you make believe you're a girl'? 'Boy: Umm. Sort of.' (The therapist thought, in retrospect, that the response may have been due to 'the interviewer's suggestion'.)[110] Green told another of his patients, Richard, aged 5, that when he grew up he would get married and 'you'll help the mommy have a little baby, and that's why it's good to be a boy and that's why it's good to have a penis – so you can help the mommy. She needs you.'[111]

The therapists made dubious claims of success. When interviewed at the age of 12, Carl (not Karl) proclaimed that he 'used to be a queer, but not anymore'.[112] But Carl seemed to adapt his speech and gesture mannerisms to the environment in which he operated, and the feminine proved rather resilient.[113] Another patient, a 7-year-old, said of his earlier cross-dressing, 'Oh, that was baby stuff. I was just mixed up. I must have thought I was a girl or something. Now I don't.' Then, in his *third* year of treatment, this young subject talked of becoming an actor, and was described by the team as 'remaining somewhat feminine in gesture and mannerisms'. Nonetheless, two years later – and at the tender age of 9 – the boy was deemed to have 'repressed the active, conscious transsexual yearnings', although he still did not like physical sports.[114]

Many historians will be horrified by a therapy that defined masculinity in terms of fantasies of beating, torturing, and raping women and which viewed the cultivation of such imaginings as progress: 'Lance wants to play murder with Barbie.'[115] We probably think it perfectly reasonable for a 6-year-old boy to say that he wants to be a girl because 'Girls don't have to play rough and get hurt.'[116] We might ridicule a regime that saw clothing colour-coordination, neatness of hair, a fascination with Cher, love of Snow White and Cinderella, a preference for Catwoman over Batman, wearing a father's baggy T-shirt, or dressing up as Mary Poppins as grounds for psychiatric treatment.[117] We may laugh at such rigidly defined notions of appropriate male and female toys – girl's toys associated with 'maternal nurturance' and boy's toys with 'masculine aggression' – and the attribution of normality to such assumptions.[118] Bryant has shown that some contemporaries were highly critical too.[119] We could question the wisdom of the four-times-a-week psychoanalysis of a 5-year-old by someone for whom 'this is the first child I have ever attempted to treat psycho-analytically'.[120] We should be sceptical of 8-year-olds who see themselves as sissies or fags, or cured queers, and of therapy that identifies early interest in 'artistic things', dressing up in girls' clothing, parental hostility to boyish aggression, motherly closeness, and female decision-making in the home as evidence of an incubating male-to-female transsexuality, an incubation that could occur very early: 'This mother had been extremely close to her infant son during the first year of life. She carried him around with her, pressed against her body almost constantly during the first 12 months – behavior described by Stoller as etiologic for male transsexualism.'[121]

The gender logic behind such interventions now appears both misguided and dangerous, especially when it involved the very young. Phyllis Burke wrote scathing accounts of the treatment of Carl and other children.[122] However, the therapists were not unaware of their highly gendered assumptions: 'While privately, one might prefer to modify society's attitudes toward crossgender behavior, in the consultation room with an unhappy youngster, one feels far more optimistic about modifying the behavior of that one child than the entire of society.'[123] Green always maintained that he was trying to alleviate the suffering of these children in an intolerant society, to 'reduce their current pain and permit them a wider range of options in the future'.[124] Yet, clearly, these therapists were reinforcing the problem rather than challenging it.

The reader may have detected a predisposition to blame women, especially mothers, in the making of a male-to-female transsexual. It was a bizarre form of Freudianism: 'I wish to expand on how the trans-sexual's mother, with her feelings of deprivation and worthlessness as a female, uses this son [*sic*] as the penis – her perfect phallus – she always wanted.'[125] Stoller encapsulated this tendency in an article called 'Transvestites' Women' that claimed that the females in the lives of male transvestites played 'an essential role . . . in the cause and maintenance of transvestism'.[126] This was achieved, Stoller argued with Catch-22-like logic, either by a mother's conscious man-hating (Stoller refers to 'malicious male-haters'), deliberately dressing her 'son' as a female and demonstrating hostility to all things male, or by her suffocating inability to separate herself from her child (what he termed the 'symbiote'), with the result that 'he' became 'feminized by his [*sic*] inability to separate his [*sic*] ego boundaries from his [*sic*] mother's body'.[127] Generations of women were culpable:

> Because of the unconscious needs generated in this mother's past, this infant son [*sic*] is fated to serve as the treasured phallus for which she has yearned. So he [*sic*] is to be the cure of the lonely, hopeless sadness instilled in her by her cold and powerful mother and rejecting father, and he [*sic*] is also to be the penis that will equalize the feeling she has had of being inferior by not being male.[128]

The substitute penis would grow up to be a 'man' who wanted a vagina! The other enabling women in a transsexual's life were the sisters, girlfriends and wives (the 'succorers') who encouraged 'his' cross-dressing.[129]

This does not mean that fathers were blameless. 'Transvestites' Women' finished by reminding its readers of the role of fathers in male feminization:

> the failure of his father to be an adequate model of masculinity with whom the child can identify when he [*sic*] needs to turn to a man, and even earlier in the boy's [*sic*] existence, the failure of his [*sic*] father to act as a shield protecting the boy [*sic*] against the urges to feminize him [*sic*] that his [*sic*] mother or sisters may have.[130]

Money was later to reject what he called Stoller's Neo-Freudian focus on mother and son in favour of a seduction of the father theory.[131] 'This is the formula: the father covertly courts his son's [sic] allegiance, in place of what he finds missing in his wife, and casts him [sic] in the role of a wife substitute, if not for the present, then for the future.'[132] Money went on to explain the collusion of the 'son' in this scenario:

> The son [sic], for his [sic] part, may solicit his [sic] father's allegiance as a formula for keeping him [sic] in the household, and for preventing a parental separation. If the father has already gone, or even if he had died, the son's [sic] gender transposition may serve to solicit his [sic] dad's miraculous return. His [sic] life becomes a living fable of the boy [sic] who will become daddy's bride, for the evidence is plentiful that a daddy can be counted on to return to the home that his wife keeps ready for him.[133]

While Stoller and his colleagues devoted most attention to the male-to-female transsexual, they also treated and reported on 'girls' who identified as male:

> In the present series of cases, the mother is almost always psycho-logically removed from the family, usually by depression, early in the girl's [sic] development. The father, while a substantial person in most regards, does not support his wife in her suffering but instead sends a substitute into the breach. This surrogate husband is the transsexual-to-be, also chosen perhaps because she [sic] strikes her parents as unfeminine in appearance from birth on. Since the family needs the child to function thus, any behavior construed as masculine is encour-aged, and feminine behavior discouraged, until the islands of masculine qualities coalesce into a cohesive identity.[134]

The logic was similar. However, whether posited as daddy's bride or mummy's husband, these theories of the parental nurturing of post-natal gender-crossing are unlikely to have resulted in any meaningful therapy.

Another approach, explored by Money and others from the late 1960s, was the use of the synthetic steroid Depo-Provera that was later to acquire some notoriety as an injected hormonal contraceptive for women. Though it was used for treatment of more extreme forms of male-to-female transvestism involving inappropriate public display, paedophilia, or self-harming activity, its advocates held out the promise of wider application.[135] Treatment, in this early usage of the drug, was to inhibit desire: the injected hormone reduced testosterone to prepubescent levels.[136] The risk, Money explained, was that this hormonal tuning could increase the transvestite's femininity (the opposite of the desired outcome) rather than inhibiting the sexual impulse to cross-dress.[137] Renée Richards claimed that Money treated her with this method and with no discernible result: 'By this time I was frantic, and even the most farfetched scheme seemed preferable to no action at all.'[138]

Behaviour Therapy

Readers may have heard of the heteronormative horrors of 1960s Anglo-American aversion therapy, when homosexuality was combatted by nausea-inducing injections and electric shock treatment.[139] However, the method was initially trialled with transvestism and transsexuality. A 1961 study, 'Behavior Therapy in a Case of Transvestism', researched and written in the UK but also published in the USA, explained the rationale for its techniques: 'The treatment is based on the hypothesis that the aberrant behavior has been acquired as a learned response, and must be abolished along the lines dictated by the laws of learning.'[140] The authors of the study claimed it as a first: 'We have been unable to find any account of this treatment applied to transvestism prior to our preliminary report.'[141] The patient was subjected to a torturous regime of oral and intravenous-induced vomiting, nausea, and headaches (apomorphine was used) while tapes and images of his transvestism were played and projected so that he would associate the cross-dressing with extreme discomfort. This continued every two hours for six days and six nights, while the patient was also given dexedrine to keep him awake. The therapists took slides of the man in various stages of dressing to represent the desire that was going to be modified, and recorded tapes of his commentary on his behaviour – 'I have now put on and am wearing a pair of ladies' panties' – that would be used 'to insure the presence of the stimulus even when the patient's eyes were closed during vomiting'.[142] The patient lay on a bed in a darkened room, and, at the moments of nausea and headache that followed the apomorphine injections, was shown the slides and played their accompanying commentaries. The attendant psychiatrists had been alert to the dangers of the process, 'the possibility of dehydration and salt depletion and even gastric hemorrhage', but reported no complications until the treatment neared its end. The planned regime of seventy-two trials was terminated at the sixty-eighth because of their patient's high temperature, elevated blood pressure, impaired coordination, and inability to converse.[143] Nonetheless, the therapy was declared to have been successful. The man was able to wear his wife's dress without arousal (a test) and was optimistic about his future. He reportedly declared his cross-dressing 'a ghastly nightmare'.[144]

Another case involving the same team and male transvestism entailed the use of faradic (shock) aversion therapy. It was considered a more accurate method than chemical stimulation because of the difficulties in timing the conditioned stimulus (the cross-dressing or its imagery) with the unconditioned response (the nausea and vomiting). Electricity was more precise.[145]

Those participating in the 1961 study had been quite clear that they were treating (male) transvestism rather than transsexuality. The transsexual, they stressed, 'believes that he [sic] is really a woman, and demands surgical conversion of his anatomic sex'.[146] But it was inevitable that the treatment would drift across categories. Psychiatrists from London's Maudsley Hospital reported in 1969 to a symposium on transsexuality that they had used electric shock aversion therapy to treat ten transvestites *and five transsexuals*. The aim was to change fantasies in the hope that this would translate into the transformation of attitudes and behaviour. Improvement was defined by the diminution or disappearance of 'deviant sexual fantasies', their severance from or minimizing of actual sexual arousal, and the absence of this fantasy work from masturbation and heterosexual intercourse. The aim was to produce 'heterosexual fantasies'.[147] The target was the transvestite and transsexual orgasm.

Although the Maudsley study appeared in Green and Money's *Transsexualism and Sex Reassignment* (1969), and claimed to treat that condition, the clinicians were only actually attempting to modify cross-dressing. The transsexuality of their transsexual patients was incidental to the aversion. The clinicians were combatting – as deviance – the desire to wear female clothing and the fetishism associated with that dress, not the conviction that the patient was a woman. It was entirely predictable that their verdict was that such treatment was more effective on transvestites than 'pronounced' transsexuals.[148]

Other attempts followed. A team of psychiatrists led by David H. Barlow at the University of Mississippi Medical Center in Jackson claimed another first in 1973: 'the first successful change of gender identity in a diagnosed transsexual'. In fact, they professed to have turned a transsexual into a homosexual and then into a heterosexual.[149] We will return to this case shortly.

It is important to reiterate that these attempts to combat what we now call transgender come from a time when endocrinological treatment, surgery, and psychotherapy had become possible for such patients to achieve their desired transition: male-to-female and female-to-male. However, the cases here ran counter to other comprehensions of transsexuality and transvestism, and to later thinking on transgender, in that they advanced 'the possibility of psychosocial intervention as an alternative to surgery in the treatment of transsexuality'.[150] The editor of the cross-dressing magazine *Turnabout*, worried in 1963 that British techniques were already being tested at Johns Hopkins, referred to aversion therapy as 'psychosurgery applied to the human soul' – a kind of 'psychic vivisection'.[151] Transsexuality, according to the logic of such therapists, was something to be resisted, rather than accommodated or facilitated: a 'deviation'. The help that was given was not for the patient to achieve a new gender identity, the dominant (though not uncontested) medical mode of thinking by the 1970s, but for them to become 'completely sexually reoriented'.[152]

Moreover, the means of achievement of 'normality' were based on highly stereotyped assumptions of gender in both measurement and treatment. Before therapy even commenced, the subject of the Jackson study, 'a 17-year-old male transsexual [*sic*]' exhibiting 'female role behavior', was measured with the 'Transsssexual Attitude Scale' which ranked 'him' (female status was never granted) according to 'his' response to a series of statements: I want to have female genitals;

I want to be a girl; I want to have long fingernails; I would like to have a large bustline; I would like to have long black hair; I want to be a female prostitute; I would like to be a waitress; I want to be a secretary; I would like to have a sex change operation; I want to have intercourse with a man.'[153] 'He' scored 32 out of a potential 40 in this measurement, indicating transsexuality. Next, 'his' arousal pattern was measured with a mechanical strain gauge that monitored penile tumescence (or lack thereof) when presented with a series of slides of naked women and men. Here the patient experienced 'virtually no arousal' to women and high sexual arousal to male imagery, 'averaging around 50% of a full erection'.[154] The final pre-treatment measure was 'his' record of sexual fantasies, including the sex of the person in that fantasy. Here, 'he' achieved a baseline score of seven 'homosexual urges' and none that were 'heterosexual'.[155]

The Jackson study was audacious in its gender modification. Actual treatment started with an attempt to replace what was identified as homosexual sexual arousal with heterosexual sexual stimulus, and the use of shock treatment as aversion therapy to counteract transsexual fantasies – techniques that will be familiar to the reader from earlier descriptions. The patient was encouraged to focus on their favourite masturbatory fantasy and then, at the height of their arousal, hetero-sexual images were introduced in a process termed 'fading'.[156] Despite eight such sessions, no change was observed. The urges and fantasies persisted. The therapists then moved to more negative conditioning, triggering electric shocks instead of heterosexual stimuli at the time of heightened desire. This too was an abject failure: 'Despite over 48 daily half-hour sessions, no changes were noted in patterns of sexual arousal nor in reports of urges or fantasies.'[157] This phase had lasted two months.

The therapists then decided, in effect, to build a man. The patient's 'gender-specific motor behavior' was recorded: their patterns of bodily deportment. Did they sit and walk like a man or a woman? Then came the treatment, modification in line with the test results – sitting, stand-ing, and walking in the appropriately masculine way. Male therapists modelled the required behaviour, mistakes were analysed and dis-cussed, progress was praised. Video playback facilitated reconsideration and adjustment. Then therapy moved to social interaction: discussing football, asking a girl to go on a date. Female research assistants were introduced to give the demonstrations verisimilitude. Finally, in this gender modification phase, therapists worked to change the patient's perceived effeminate speech patterns: 'a more relaxed speech with some slurring was a therapeutic goal'.[158] Of course, the content of that speech had to be appropriately masculine: 'I like being a boy' or 'A good looking woman turns me on.'[159]

It is unclear how long the gender behaviour phase of the treatment took. The therapists mention thirty sessions over two months, but this may refer only to the treatment before voice modification. They had created someone who looked and acted like a boy (the subject was only 17). However, mind had not matched body. The patient 'felt like a girl', scored 32 on the transsexual attitude scale, failed to achieve erection when presented with images of women, and still experienced sexual urges and fantasies that the therapists persisted in describing as homosexual.[160]

They then set about tackling those impulses with 'Fantasy Training' to wean 'him' away from imagining 'himself' as a woman having sex with a man, towards more 'gender-appropriate fantasies in which the patient, as a man, had intercourse with a woman'.[161] The transsexual chose four *Playboy* images that were 'the least unattractive' and was asked to imagine sexual contact with the pictured woman.[162] The maintenance of the fantasy was measured by the subject's symbolically raised index finger (indicating that the image was clear) and this sustaining rewarded by praise, but also with another picture of something that 'the patient had previously chosen as very pleasant'.[163] (The fact that these pictures were of food or animals seems to risk introducing a somewhat bizarre complexity into this sexual modification, but we will not pursue that thought.) These sessions were held daily and presided over by a female therapist who urged encouragement, and who also 'would suggest pertinent behavior with which to enrich the fantasy such as explicit details of foreplay'.[164] These thirty-four sessions took a total of two months and resulted in some ambiguous results, although the authors of the study did not see them as such. The patient had lost their transsexuality as defined by the transsexual attitude scale (scoring a mere 4 out of 40) and claimed no gender role reversal in their sexual fantasies. The psychiatrists were encouraged by the fact that women as well as men were recorded in these fantasies (and that those women were not the patient). However, this sat strangely with the absence of any measure of penile response to women (they were in 'his' imaginings but 'he' did not desire them) and by the erections always attained when faced with 'male stimuli'. They concluded – erroneously, surely – that their subject was now 'diagnostically a homosexual'.[165] And they set about converting this new homosexual – whose transsexuality had been seen as homosexuality in the first place!

The final task was to change the patient's 'Sexual Arousal Patterns'. Note that the previous phase had been directed at fantasy change and that now the emphasis was on increasing heterosexual arousal; the distinction between the two is significant. Treatment reverted to the classical conditioning and aversion therapy carried out at the start of what had become a therapeutic marathon. This time there was claimed success. Penile circumference changed towards erection in the presence of female erotic stimuli. Twenty sessions of aversion therapy over two months reduced homosexual arousal. A year after treatment, the transsexual score was zero, and heterosexual and homosexual arousal were gauged, respectively, at 55 and 15 per cent.

A follow-up study published in 1979 by Barlow, now at Brown University, indicated that this treatment had been replicated on two other patients to decrease their 'homosexual and transsexual arousal'.[166] He appears to have upped the ante in the final phase of heterosexual arousal reinforcement by introducing a 'final orgasmic reconditioning phase', with actual masturbation in the laboratory to reinforce heterosexual fantasies.[167] Barlow billed his teams' modifications as 'apparently successful'.[168] However, a critic could easily question both the initial diagnosis and the claimed changes in identity, gendered behaviour, and sexual arousal. Three-year follow-up tests with penile response measurements indicated that the original patient still experienced considerable homosexual attraction and had not had intercourse with 'his' girlfriend.[169] One of the other patients lived as a gay male after treatment, reporting that 'heterosexual activity moved too slowly and that there were too many preliminaries', while a third engaged in homosexual casual sex.[170] If there was any identity change for those transsexuals, it was towards homosexuality, rather than the heterosexuality intended by the clinicians.

But what was homosexuality in that transsexual moment? For one of the authors of *Transsexualism and Sex Reassignment* – and for most male-to-female transsexuals – 'The transsexual . . . has sexual relations with members of the same anatomic sex because he feels he is feminine and heterosexual . . . He considers sexual relations with women to be homosexual.'[171] A male-to-female transsexual, a trans woman, would have seen desire for men as heterosexual (because she was really a woman), but for Barlow and his colleagues such longings were homosexual (because they saw 'him' as a man). In the mirror world of this particular piece of psychiatry, heterosexuality was the targeted patient desire for sex with women, which is lesbianism in transsexuality and transgender. When Barlow claimed that 'patterns of sexual arousal were changed from homosexual to heterosexual', he was correct in transgender terms – though, of course, he did not mean it in that sense.[172]

Conditioning had its perils, and not just the complications of linking arousal to imagining food or furry animals, or the later admissions of some patients that they had lied about the effectiveness of their treatment, and the disclosures of some nurses who had colluded.[173] When, in orgasmic reconditioning, the masturbating patient used an inappropriate image or memory to arrive at the point of orgasm and then shifted to the appropriate stimulus as they ejaculated, what was the guarantee that the original, more arousing, fantasy had not been invoked?[174] Therapists were aware that the aversive response could fasten on aspects peripheral to the target – to shoes worn, the photographic process – or even to the therapy itself: 'the conditioned avoidance response may result in the patient never seeking treatment again for his disorder'.[175] The authors of the transvestite studies knew that, because of the inaccuracy of timing with chemical conditioning, they might have been reinforcing the very behaviour that they wanted to extinguish.[176] Those who applied the faradic treatment knew that the sheer number of times their patient dressed and undressed (400 times in six days) may in itself have explained the loss of appeal in donning female dress.[177] Alternatively, there was the spectre of sexual pleasure from the electric shock: 'Thus, far from the electrical shock being averting, it might become part of the normal fantasy situation.'[178]

The greatest potential fantasy complication of all occurred with the images used in fading. Tantalizingly, when colour photographs of a naked male were replaced by those of a naked female, they represented the very ideal of the bodily transformation of a trans woman, as breasts and hair grew, the penis faded, and female genitals replaced the offending member – though this irony appears to have been lost on the psychiatrist therapists.[179] See Illustration 21. (Incidentally, the image changed race, too. The naked male is white and the naked female is black!) Unsurprisingly, then, when in 1979 the British clinical psychologist Harry Brierley summarized such treatments in a handbook for counsellors, psychiatrists, and psychologists, he described them as 'somewhat disappointing', 'a multiplicity of shots in the dark'.[180]

Definition/Diagnosis

Another issue in the early years of transsexuality was the question of its actual identification. 'Although some clinicians ... still maintain the idea that a group of pure transsexuals exist for whom SRS [sex-reassignment surgery] ought to be prescribed', Lothstein wrote in 1977, with Benjamin and Stoller in mind, 'the fact is that patients requesting such surgery constitute a diverse group of individuals suffering from a broad range of gender disorders'.[181] In *Female-to-Male Transsexualism*, he took the case of 'Tina/Tim', an archetypal female-to-male transsexual: 'I have no doubt that Tina would have been judged as a "true" or "primary" transsexual, if not as a good candidate for surgery at most gender clinics.'[182] And yet Lothstein was able to cast considerable doubt on the certainty of the initial diagnosis, demonstrating a weakness of diagnosis and reappraising treatment, despite the involvement of therapists with psychiatric and psychological training – indeed, their testing facilitated his revisiting of the treatment process. The point was that many transsexuals were being treated by those with no such expertise and where the potential for error was far greater.[183]

Jon Meyer, a psychoanalyst at Johns Hopkins, argued that the term 'transsexualism' had become a victim of its own rapid diagnostic acceptance. Its 'reification' disguised huge clinical variation: transvestism, masochism, conflicted homosexuality, polymorphous perversity, schizophrenia, and true transsexualism.[184] The 'self-designated transsexuals' that he had treated fell into one or more of those numerous categories.[185] They included 'aging transvestites', most of whom

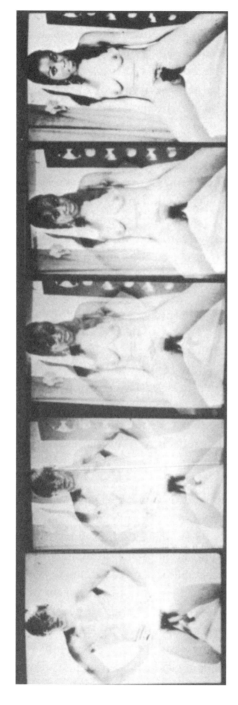

21 *Images (originally in colour) projected during behavioural conditioning in the treatment of homosexuals and transsexuals, 1973.*

were taking hormones, but only one of whom was granted surgery because he considered them 'poor candidates for sex reassignment'.[186] They included those he designated as homosexual – both male and female – who were denied surgery because 'sex reassignment surgery is sought by these patients seemingly to lend quasi-biological and medical rationalization to what is viewed otherwise – by the patient – as a perversion'.[187] And they included the polymorphously perverse, whose 'erotic behavior is dependent on opportunity and convenience, rather than internal preference'. 'It seems unreasonable to view them as having made a commitment to any one social or sexual role.'[188] Accordingly, Meyer suggested a more comprehensive term, 'gender dysphoria syndrome', for such cases, and urged that transsexual and its derivatives be reserved for those who had actually undergone surgical reconstruction ('It could be used much as the term "amputee" to describe a postoperative fact').[189] He also questioned whether transsexualism represented 'a true *reversal* of core gender identity', given that the cases that he dealt with 'indicate more ambivalence and ambiguity of gender than fixed reversal'.[190] This was an interesting anticipation of the later transgender turn.

It is evident from a close reading of Meyer's 1974 report that Johns Hopkins was overseeing the reassignments of patients whom he classified as other than archetypically transsexual. While none of the five so-defined schizoid or six homosexual patients received genital surgery, one of the ten fetishistic transvestites, one of the four masochists, and two of the three polymorphously perverse underwent sex reassignment.[191]

A report in the same year from the Department of Surgery at the Stanford University School of Medicine, drawing on 769 patients and 74 operations, charted a similar situation. The 'Stanford experience', though indicating a cautious approach in the fact that less than 10 per cent of patients were actually operated on, made it evident that transsexualism as a category had become a catch-all for people best defined in other gender dysphoric ways. While some of the discussed misdiagnosis was either historical or the problem of other clinics – 'many non-transsexuals, having read [Harry] Benjamin's book, presented themselves to physicians as classic Benjaminian transsexuals' – it is obvious that Stanford was providing surgery for male-to-female patients diagnosed as effeminate homosexuals and transvestites, as well as those considered to be true transsexuals. More than a third of male-to-female patients thus operated on were classified as effeminate homosexuals rather than transsexuals.[192] And if it is true that the Stanford experience claimed to refuse surgery to those patients deemed psychotic, neurotic, or sociopathic, it is also evident that its medical experts began their report with a description of their patients as those 'whose determined quest for rehabilitation *via* surgery is an almost psychopathic drive'.[193] It must have been hard to separate the alleged psychopaths from the alleged psychopaths.

Stanford had hosted an international symposium on gender dysphoria in 1973, and the proceedings of that gathering of gender identity experts certainly give pause. A Stanford representative, Norman Fisk, admitted to an early 'inexperience and naiveté' in their programme. A 'great emphasis was placed upon attempting to exclusively treat only classical or "textbook cases" of transsexualism': a life history of feeling that one belonged to the other sex; non-erotic cross-dressing; an aversion to homosexuality; and a conviction that their mere request for transsexual surgery confirmed their transsexual status. Unsurprisingly, the Stanford experts were offered a string of identical textbook cases – 'far too many patients presented a pat, almost rehearsed history, and seemingly were well-versed in precisely what they should or should not say or reveal'.[194] In hindsight, Fisk considered this less a case of deliberate fabrication than a pattern of retrospectively rewriting sexual histories to conform to the expected transsexual stereotype: 'Here, the patient quite subtly alters, shades, rationalizes, denies, represses, forgets, etc., in a compelling rush to embrace the diagnosis of transsexualism.'[195]

As the Stanford team became more experienced and more inclusive and flexible in their gender conceptualizing, shifting their diagnostic focus from transsexualism to gender dysphoria, and considering surgery for dysphoric patients not in the transsexual category – what Fisk termed a 'liberalization' – they found, again unsurprisingly, that the histories they were given were more 'honest, open, and candid'.[196]

Definitions varied and the most fixed classification disguised uncertainty. Identification of transsexuality was beset with substantial diagnostic complications.

Assessment

Another major issue was assessing the success or otherwise – of treatment. Stoller, we saw, was scandalized by the paucity of follow-up studies of (and this was another criticism) an unknown number of possible patients. Kubie was concerned about the reliability, or even feasibility, of such research: 'How does one obtain the objectivity necessary to evaluate the concepts, the practices and consequences of sex-change operations? The methodological problems are extraordinarily difficult.'[197]

The post-surgical picture from the University of Minnesota's 1960s Transsexual Research project was not encouraging. Twenty-five male-to-female patients were assessed five years after surgery (most of the operations occurred in 1968) and reported on as part of the research project that had provided free surgery in return for the subjects' participation (the practice changed shortly afterwards). The author, Donald Hastings, summarized patient satisfaction as good, 'Not excellent, but good'.[198] However, many of the extracted patient case notes are more negative in terms of patient progress, surgical success, and psychiatric attitudes: 'Sociopath'; 'Sociopath, public prostitute, marginal IQ'; 'Vaginal repair NYC 12/69'; 'Probably gave lues [syphilis] to husband of another transsexual'; 'attempted suicide by shooting self in abdomen in front of boy who jilted her'; 'Calls suicide "hot line" to complain how fast whiskers grow'; 'Alcoholic . . . serious suicide attempt'; 'Developed severe infectious hepatitis; long period of sexual maladjustment due to surgical complications'; 'Severe sociopath'.[199] Hastings concluded that the 'surgical route' was the only, rather than the best, course, and he was candid in his admission that patients detected his 'disapproval'.[200] It was hardly a ringing endorsement of the treatment of transsexuality.

Thomas Kando, who published a sociological study of the same post-operative Minneapolis patients, seemed to define the success of their transitions in terms of their surface femininity and his own personal attraction. He wrote of contrasting fates. The fortunate were those of 'near perfection', 'profoundly attractive', the 'tall beautiful' blondes, with short miniskirts and a 'girlish shyness', 'very attractive', 'exquisitely feminine', 'highly feminine', young women.[201] 'The fact that she may have been more interested in me as a prospective date than a sociologist may have been the healthiest sign of all.'[202] The less successful, categorized as simply 'odd', were older, larger, 'grotesque', and manlike: 'it was difficult to visualize how they would surmount the barriers that still separated them from any semblance of woman-hood'.[203] Yet the most successful, according to these somewhat shallow criteria, were not without post-surgical problems. One tall, beautiful blonde talked of attempted suicide and corrective surgery and said that she had 'not been able to function as a woman, yet'.[204] Kando said that most of the patients he interviewed had had to return to hospital for corrective surgery, and that one woman was wearing leg braces 'due to temporary paralysis resulting from the operation' and had been back four times for corrective surgery.[205]

Wardell Pomeroy, a former member of the famous Alfred Kinsey team at the Institute for Sex Research at the University of Indiana, was similarly ambivalent in his 1969 report on the sexual lives of twenty-five pre- and post-operative trans women whose histories he had collected.[206] It was a small study: only eight of the eleven patients who had received gender reassignment were interviewed about their lives after surgery. And it was a bleak survey, despite Pomeroy's conclusion that the 'conversion operations were beneficial'.[207] The discussed cases included women with little or no experience of intercourse, masturbation, or orgasm (Cases 1, 7, and 8), and one with a vaginal opening too small for intercourse (Case 2). Pomeroy thought them people with 'rather low rates of overt sexual behavior but a very great fantasy life'.[208] He was sceptical about the orgasms of the claimed orgasmic: 'it is doubtful from her description that the patient has ever had an orgasm since the operation'.[209] Although Pomeroy was supportive of facilitating such transitions and considered his subjects' lives improved, this may merely have indicated the true desperation of their former condition.

Green, a psychiatrist, wanted to impart realism into therapy and surgical expectations: 'Many transsexuals have led lonely, isolated lives prior to surgery. They optimistically look forward to reassignment as a rebirth. They may harbor unrealistic expectations of an immediately blissful life, exciting and romance-filled.'[210] Thus, trans women should not expect to achieve 'idealized female proportions' from taking oestrogen. They should be aware of the costs and time involved in electrolysis. They were informed of vaginal vagaries.[211] Trans men should not assume breast reduction from testosterone treatment, and might anticipate that the desired growth in their bodily and facial hair may also be accompanied by acne. They could anticipate clitoral growth but not a 'phallus of penile dimensions'. As for constructive surgery: 'Female transsexuals [*sic*] should know that construction of a penis is still in an essentially experimental stage.'[212]

Experienced practitioners at Johns Hopkins were very indecisive about their patients' emotional prognosis after reconstructive surgery, but quite sure that they were not working any miracles, informing readers of the specialist *Plastic and Reconstructive Surgery* that the male-to-female 'reassignee . . . is not – and never will be – a real girl but is, at best, a convincing simulated female. Such an adjustment cannot compensate for the tragedy of having lost all chance to be male and of having, in the final analysis, no way to be really female.'[213]

Some were candid in their assessment of surgical complications. Stanford's 1974 report stated that almost half of the male-to-female post-operatives and a quarter of the female-to-males suffered complications. Recto-vaginal fistulas, narrowing or closing of the vagina and/or urethra, blood loss, and infection were among the problems listed (it is not certain whether 'Excessive emotional attachment to surgeon' and 'Desire to shoot genitals of surgeon, with a shotgun' were complications for the patient or the doctor).[214] Those operated on elsewhere but coming to Stanford for treatment provide hints of a potentially far worse situation, including many trans women with 'inadequate . . . or no vaginas'.[215] A disconcerting feature of the Stanford experience is that it was an experiment: 'In our program, transsexualism has been questioned as a disease, and its surgical treatment has been investigated in a clinical trial.' The verdict in 1974, for a programme started in 1967, was that 'Surgery is *not proven* to be the treatment for the transsexual condition.'[216]

Stoller provided a somewhat pessimistic summary of what he termed the 'transsexual experiment' in a collection of his late 1960s and early 1970s writings – and this was from an acknowledged pioneer. He suggested that, even after surgery, and despite bodily reconciliation with what was perceived as their true identity, trans women were still psychologically – and therefore sexually – troubled by the 'boy' that still lingered inside them. 'Each sex act was not only an erotic experience but also a test of the success of her body transformation, and since her partner's penis was in where "he" (her boyhood) still lived, the patient could never relax into the safety of a complete sense of femaleness.'[217] As one patient pointed out, 'I mean, after all, I have still got the same penis. It's just differently arranged that's all.'[218] Stoller wrote of the 'removal of the unwanted male sex' failing to extirpate a deeper sense of maleness.[219] Yet it was even more problematic given that the phallus was inverted as a vagina rather than removed; the transsexual's partner's penis was actually inside her penis. The point is that sexual adjustment after surgery was difficult – what the earlier-quoted patient referred to as a 'pussy bummer'.[220]

Lothstein also raised questions about the claimed accomplishments of reassignment surgery. He took the case of a trans man, whose surgery was proclaimed successful and who was deemed to have adjusted satisfactorily. Yet when Lothstein explored the test results more deeply and arranged for follow-up testing, he discovered that Tim had lied about his satisfaction with his penis because he had not wanted to disappoint his therapist. He felt socially isolated and seemed regretful about his surgical path. 'In effect', wrote Lothstein, 'her [sic] real penis was not as powerful as her [sic] imaginary one'.[221]

Conclusion

It is clear, then, that the history of transsexuality should include this extensive critique. The 1978 selected proceedings of the Fourth International Conference on Gender Identity, published in the journal *Archives of Sexual Behavior*, provide a good snapshot of the mood as the 1980s approached. One expert contributor observed 'a proclivity for the creation of self-fulfilling, self-validating, reinforcing cycles'. It was 'a theorist's paradise – for in the absence of any objective criteria, the stage is set for a happy collusion of fantasies between therapist and patient'.[222] The transvestite Virginia Prince referred to sex-reassignment surgery as a 'communicable disease', easily spread among the 'susceptible . . . transvestite and drag queen population'.[223] A therapist from the University of Pennsylvania warned that they had to deal with sexual stereotypes 'in a time when sexual stereotyping is not generally applauded'. 'We cannot act as if we are somehow participating in the nostalgia of these persons using techniques of the 1970s to create a caricature of the 1940s movie version of a woman.'[224] In an effort to impose some reality on the situation, he sent some of his candidates to feminist group sessions.[225]

Nor were the symposium's reports of surgical advances without their downside. The Stanford-constructed penis did not have a urethra, was not capable of erection (without a removable prosthesis or unless scarring and fibrosis provided a semi-erect state), and its recipient still depended on his clitoris for sexual stimulation. Post-surgical complications were common. Testicular implants (Stanford experimented with lead in silicone!) seemed especially problematic: of ten procedures, six became infected and three of those patients lost their implants.[226] Urologists and plastic surgeons from Cook County Hospital, Chicago, reported an increase in post-surgical complications, mostly, they stressed, among those who had had surgery elsewhere. Most common was a closing of the vagina, often the result of inadequate post-operative care, but more surprising was the problem with penile stumps: 'These stumps of corpora cavernosa became engorged and painful during coitus and forbid the patient from further attempts at intercourse.'[227] Photographs of disfigured genitalia provided visual confirmation of this rather grim verdict.[228]

There are several references to patients who shopped around for treatment – one of the problems for follow-up studies was that their subjects had moved on.[229] Those at the larger gender clinics knew that, since the vast majority of their patients were turned down for gender reassignment, it was inevitable that they went elsewhere ('surgeon shopping').[230] Charles Ihlenfeld, in private practice, said that he was sure that many of those he treated were 'fugitives from university gender identity groups', and was clearly disillusioned with his clientele: 'after six years and several hundred patients ... I feel manipulated and blackmailed to the point where I seriously question whether patients really should be seen in this way'.[231] The spokesperson for the Stanford University Gender Dysphoria Program wrote that 'many' of their patients grew impatient with the cautiousness of their approach and sought a solution elsewhere, with, he thought, damaging results. 'The human wreckage in this group is astonishing testimony that sex conversion at inappropriate times and for inappropriate candidates is a devastatingly destructive procedure.'[232] There were also the inexperienced practitioners eager to meet the market demand. The doctor from the Pennsylvania Hospital at the University of Pennsylvania said that there were physicians who 'will prescribe hormones to virtually anyone who requests them', and claimed an 'estimated 200 self-diagnosed male-to-female transsexual candidates in the Philadelphia area alone who have received estrogen therapy for one or two years without *ever* having seen a psychiatrist'.[233] Other doctors discussed the 'tremendous black market' in liquid silicone injections by non-physicians.[234]

There is an argument, then, for a counter-history. One major strand, we have seen, and that Meyerowitz has identified, is the psychoanalyst's misgivings about the efficacy – indeed, desirability – of transsexual surgery.[235] Johns Hopkins's Meyer told the *New York Times* that his 'personal feeling' was that surgery 'was not a proper treatment for a psychiatric disorder, and it's clear to me that these patients have severe psychological problems that don't go away following surgery'.[236] His boss, Paul McHugh, responsible for closing down transgender surgery at Johns Hopkins, stated quite bluntly that 'Hopkins was fundamentally cooperating with a mental illness.'[237] He later gave the facilitation of 'sex reassignment surgery' as an example of 'the power of cultural fashion to lead psychiatric thought and practice off in false, even disastrous, directions'.[238] However, the critique was evident on an impressive variety of fronts, including representatives of most of those involved in the transsexual experience. Psychiatrists (recall Lothstein) questioned the diagnoses of other psychiatrists. Psychologists criticized each other. Suzanne Kessler and Wendy McKenna wrote of clinicians who used their own sexual interest and concepts of female beauty to decide the legitimacy of their patients' claims to be trans women.[239]

Sociologists critiqued the whole phenomenon. Dwight B. Billings and Thomas Urban, two social scientists familiar with the medical literature, who had interviewed both patients and practitioners in the USA, and one of whom had been a two-year participant observer in a clinic, wrote in 1982 that 'There is hardly a more dramatic instance of contemporary professional authority than so-called "sex-change" surgery . . . transsexualism . . . *only* exists in and through medical practice'.[240] Billings and Urban stressed that 'transsexualism' flattened out the complexity and variety of 'gender role distress'.[241] They were damning of the attitudes of some of the medical experts whom they encountered, one reputedly proclaiming, 'We're not taking Puerto Ricans any more; they don't look like transsexuals. They look like fags.'[242] They also stressed that transsexuals told their therapists exactly what they wanted to hear, providing the rehearsed histories needed for surgical permission.[243] They concluded that 'Transsexual therapy, legitimated by the terminology of disease, pushed patients toward an alluring world of artificial vaginas and penises rather than toward self-understanding and sexual politics.'[244]

As the sociologists' comments remind us, there was repeated disapproval of the patients. Meyer claimed that none of his trans patients had been offered formal psychoanalysis, 'since their ego strengths were considered insufficient for the task'.[245] A paper presented at a meeting of the International Gender Dysphoria Association in 1981 alleged multiple cases of fabricated transsexual patient sexual histories, and even 'the hiring of actors to impersonate family members'.[246] Catherine Millot thought that the female-to-male transsexuals she had encountered hovered on the border of hysteria and delusion: 'Some dream that one day it will be possible to transplant the penises of dead men. They madly place their hopes in the possibility of erection, and even procreation. For them, there are no limits to the power of science. It's simply a question of time.'[247] Even transsexual autobiographies contained mockery of those considered less-than-ideal exemplars of a presumably shared condition. Richards wrote disparagingly of the occupants of Benjamin's waiting room: 'I surveyed the room and found it full of creatures who were neither fish nor fowl . . . A lot of them looked as if they had lost their senses.'[248]

Romaine Atura jumped off a New York building in 1977 after regretting her surgery. She had 'progressed along the way from effeminate homosexual, to transvestite, to drag queen to finally transsexual'. However, she began to question the logic of her transition:

She was aware of how she had been coached by her friends and herself, seeking a final answer, as to say what to get over during the interview. She was aware that she had been spouting phrases that she had read, or heard, about transsexuals. She was confused as to whether or not she had been sincere or merely reading a script.[249]

There was the alleged peer pressure of the 'drag bar'. 'In her own circle of friends, the sex change was the peak of the social order.' She was living as a woman and had breasts, so the next logical step was to acquire a vagina. When she got it, she could not feel anything with it and wanted her penis back. The article outlining Atura's sad predicament was in a publication aimed more at the transvestite than the transsexual, but indicated that reservations about the surgical process extended beyond a medical elite.[250] There was no smooth path to transsexual acceptance – for any of those involved.

The Transgender Turn

Introduction

Nick Krieger's perceptively honest account of his journey from boyish lesbian to a trans masculine identity, 'outside the binary', was a conscious attempt to 'create a new transgender paradigm that went beyond being "born in the wrong body"', as he explained in the new trans text *Trans Bodies, Trans Selves*. 'I wanted to show my experimentation and uncertainty, my quest to reinvent my body rather than fix a problem.'[1] Krieger, in 2014 at least, considered themself as 'being neither man nor woman, but both'.[2] Hence their account is different to so many of the transsexual autobiographies referred to in the preceding chapters of this book. 'To the best of my recollections, being a girl wasn't anything I'd ever questioned for nearly thirty years . . . If the proof that a person was transgender came in the form of the long-sustained narrative, the history of always knowing, then I was in the clear.'[3] Krieger's decision to bind, pack, and (eventually) embark on top surgery was more in terms of bodily aesthetics than gender affirmation, and they consciously edited out descriptions of the immediate results of that surgery in an effort to avoid a 'trans narrative cliché'.[4] The fact that Krieger was one of the trans profiles in *Trans Bodies, Trans Selves* reflects that handbook's embracing of trans diversity.

Similarly, Bobby Jean Noble, in 2006, had pondered using the word 'grafting' rather than transition to describe their relation with their body after top surgery. 'My *gender* now looks different from the one I grew up with but my body is, paradoxically, almost still the same.' 'The trope of grafting', Noble wrote, 'allows me to articulate the paradox signalled by "I am a lesbian man" or "I am a guy who is half lesbian"'.[5] Krieger and Noble were a product of what has been called the transgender turn. This chapter examines the complexities of this period.

The Turn

Transgender emerged as an all-encompassing term in the 1990s to include a wide range of articulated genders. It comprised the transvestism and transsexuality discussed earlier, and (obviously) transgender itself, but also a series of other forms of gender (and sexual) expression, including an assortment of bodily transformations and an array of male and female cross-dressing. The young Susan Stryker, who, as we discussed in a previous chapter, felt out of sync with the category transsexual, discovered affinity with a 'new word getting tossed around' in San Francisco in 1990: transgender. 'I felt it fit me, and created a bit of distance between the old medical mindset associated with "transsexual" and the bohemian life I was living.'[6] It is important to stress – in the face of the transsexual insistence on privileging gender over sexuality – that the two were firmly linked in the transgender turn. One of the manifestoes of transgender, Kate Bornstein's *Gender Outlaw* (1995), began with an explication of 'transgender style': 'I identify as neither male nor female, and now that my [female] lover is going through his gender change, it turns out I'm neither straight nor gay.'[7]

'We are all transsexuals', Jean Baudrillard wrote in 1993, both confirming a transgender turn and simultaneously denying it with his use of the word 'transsexual'.[8] The actual word 'transgender' was used before the 1990s, and is generally attributed to the cross-dressing activist Virginia Prince and *Transvestia* magazine in the late 1970s and early 1980s.[9] Certainly, Prince referred to 'transgenderal' in both 1969 and 1970, claiming that it was the best word to describe someone like her who had had electrolysis, taken hormones, and who lived 'as a woman full time', but had not had, and did not seek, gender-affirming surgery.[10]

Intriguingly, Prince's biographer, the gender researcher Richard Docter, used 'transgenderism' in 1988 to describe 'full-time living in the cross-gender role in the absence of sexual reassignment surgery, with oscillation, however rare, back and forth from one gender role to the other'.[11] It was not quite the umbrella term that it soon became.

The gay Italian communist, Mario Mieli, could have done with the word 'transgender' in 1980, when he was attempting to distinguish between what he called 'transsexuality', indicating a basic polymorphous sexuality (trans-sexuality), carried within every individual, and the medicalized gender identity, transsexuality (as in the transsexual moment). With the latter, society forces people into what Mieli called a 'monosexuality, seeking to identify with a . . . "normal" gender opposite to their genital definition . . . Society induces these manifest transsexuals to feel monosexual and to conceal their real hermaphrodism [*sic*].' When Mieli referred to a revolutionary 'transsexual' future, with a 'liberated Eros', where everyone recognized the 'presence of the other sex' within themselves, what he really had in mind – in the absence of the word itself – was transgender.[12]

The transgender *turn* emerged in the late 1980s.[13] It arose out of a combination of dissatisfaction with what was perceived to be the inflexibility of the university gender clinics (the medical model of transsexuality with its 'heterosexual graduates'), and the growth of private practices catering for a perceived market.[14] Docter claimed in 1988 that it had become relatively easy (with money) to obtain untrained counselling, illicit hormones, and minimally monitored surgery. 'Although there are clear standards recommended for selection of transsexual applicants, it would appear possible to arrange for such surgery in the United States virtually in the absence of any licensed professionals.'[15] But the turn was also a result of increasing transgender community organization. The strands are hard to separate. Trans accounts stress community agency. For Stryker and Paisley Currah, 'the category "transgender" represented a resistance to medicalization, to pathologization, and to the . . . medico-legal-psychiatric institutions'.[16] '[T]he transgender model arose not from the medical community . . . but from the transgender community', wrote Dallas Denny, a participant in the turn.[17] In fact, the trans woman Holly Boswell advocated 'The Transgender Alternative' in

the pages of *Chrysalis Quarterly* in 1991, an activist transgender magazine edited by Denny.[18]

The titles of a group of 1990s publications encapsulate the shift in
mood. The then butch lesbian Leslie Feinberg published the pamphlet
Transgender Liberation: A Movement Whose Time Has Come in 1992;
transgender was a term to 'express the wide range of "gender outlaws":
transvestites, transsexuals, drag queens and drag kings, cross-dressers,
bull-daggers, stone butches, androgynes, diesel dykes or berdache'.[19]
Bornstein, in 1995, was conscious of standing at a turning point, 'the
beginning of a movement', a time for new narratives and abandon-
ment of the old 'tales of women trapped in the bodies of men or men
pining away in the bodies of women', of those who exchanged one 'false
gender' for another.[20] Gordene Olga MacKenzie's *Transgender Nation*
(1994) was critical of the medical model and the privileging of surgery,
arguing that 'transsexualism is moving away from being considered a
psychological "disorder" that is treatable with surgery and hormones
to a grass roots civil rights movement'.[21] Zachary Nataf's pamphlet
Lesbians Talk Transgender (1996) contained critiques of transsexual narratives and the role of the medical and psychiatric establishment, as well
as advocacy of pansexuality.[22]

By the end of the 1990s, then, Boswell could look back on what
she termed the transgender paradigm shift: 'Never before have we
had so many options . . . We are choosing to define ourselves outside
of our cultures, and virtually outside of the very system of gender
as we have known it. Transgendered people are redefining gender.'[23]
Photographers encapsulated the turn: Mariette Pathy Allen with her
portraits of varied American transgender lives (including a picture
of Holly Boswell); Del LaGrace Volcano, with his more challenging
imagery of queer lives and trans masculinity.[24]

The transgender turn did not operate in a vacuum. It was part of a
wider cultural shift, reflected in – and influenced by – Judith Butler's
now classic theoretical destabilizing of sex and gender, a pivotal moment
in the 1990s and in queer studies generally. Such thinking and practice
threw binary certainties into doubt, questioning not only any direct and
necessary correspondence between sex and gender, but arguing that sex
itself was a gendered category. Gender, in that emerging conceptual
framework, was often considered as performance, 'a corporeal style'.[25]

The implications of this destabilizing were immense – and with clear implications for trans identities and sensibilities. 'If sex does not limit gender', Butler wrote,

> then perhaps there are genders, ways of culturally interpreting the sexed body, that are in no way restricted by the apparent duality of sex. Consider the further consequence that if gender is something that one becomes – but can never be – then gender is itself a kind of becoming or activity . . . an incessant and repeated action . . . that can potentially proliferate beyond the binary limits imposed by the apparent binary of sex.[26]

The non-fixity of gender was demonstrated in various ways – the possibility, as Butler has expressed it (referring to the fiction of Monique Wittig), of becoming a being 'whom neither *man* nor *woman* truly describes'.[27] For the transgender turn was part of 1990s postmodern, 'Fin de siècle, fin de sexe', with its figures of cyber fiction, literary invention, and film and video, performance artists, surgically transformed 'trans-intersexuals', and a range of practitioners of queer sex such as leatherdyke boys and leatherdyke daddies.[28] Richard Ekins and Dave King's book *Blending Genders* (1996) is a product of that period, with its concept of male femaling (bodily, erotic, and gender) and Stephen Whittle's iconic piece 'Gender Fucking', featuring the trans man Loren Cameron's powerful image of naked masculinity, where, in the absence of a penis, Cameron is the phallus.[29]

On a totally different register, the shift to transgender was reinforced by the Johns Hopkins Gender Identity Clinic's termination of trans-sexual surgery after the 1979 critical report by Jon Meyer and Donna Reter (discussed in Chapter 4), although we should neither exaggerate this impact nor equate surgery with other treatment.[30] Hopkins still oversaw transgender patients and might recommend surgery elsewhere. The programme at Stanford continued as the (unaffiliated) Gender Dysphoria Program at Palo Alto.[31] When a scholarly survey of surgical reassignment protocols was published in 1995, Palo Alto, the University of Chicago Medical Center, The Gender Identity Clinic, Case Western Reserve, Johns Hopkins, The Gender Dysphoria Program at the University of Minnesota, The Gender Dysphoria Program of Orange

County, San Juan Capistrano, California, The Gender Identity Clinic of New England, Manchester, Connecticut, and The Institute for Gender Study and Treatment, Arvada, Colorado, were among those able to comment on their protocols, so they were obviously offering treatment, if not always performing surgery themselves.[32]

The transition was also affected by increasing medical usage of the terms 'gender dysphoria' and 'gender identity disorder', which replaced transsexualism (DSM-III (1980)) in the American Psychiatric Association's *Diagnostic and Statistical Manuals*, moving from Gender Identity Disorder (DSM-IV (1994) and DSM-IV-TR (2000)) to Gender Dysphoria in the current DSM-5 (2013).[33] John W. Barnhill, Professor of Clinical Psychiatry at Weill Cornell Medical College, has recognized the fine line between the pragmatic ability to use a purported clinical condition to justify insurance support for gender surgery or hormonal therapy (on the one hand) and simple stigmatization (on the other), but claimed that DSM-5 had moved 'further in the direction of depathologizing discordant gender identity by developing a new diagnosis, gender dysphoria, which emphasizes clinically significant distress or dysfunction, along with the discordance'.[34] He was wrong about the newness of the diagnosis: Betty W. Steiner's *Gender Dysphoria: Development, Research, Management* (1985) had been published some thirty years earlier.[35] He may also have exaggerated the depathologizing. After all, his words were expressed in a book of DSM clinical cases, and the US-edited 2015 edition of the *Oxford Textbook of Psychopathology*, aimed at psychology and psychiatric professionals and students, and presumably reflecting recent medical thinking, consigns gender dysphoria to a chapter dealing with necrophilia, bestiality, biastophilia (erotic interest in rape), and hypersexuality.[36] As Austin Johnson has pointed out, while DSM-5 shifted the focus from gender-variant identity itself as a problem to the possible distress associated with that condition (Barnhill's argument), it still presented a 'psychiatric diagnosis' for gender variance and tended to 'overemphasize the importance of medical intervention'.[37] Zowie Davy has likewise referred to semantic changes and stigmatization.[38] Arlene Istar Lev has called DSM-5's category 'Two Steps Forward, One Step Back'.[39] It might also seem perverse to credit the DSM a role in the rise of transgender when the manuals studiously avoided any mention of the word until DSM-5.[40]

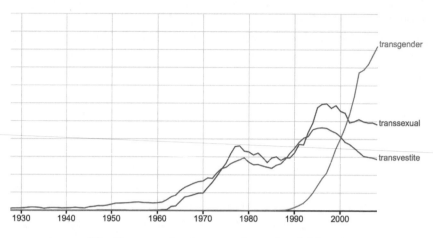

22   *Transsexual and transgender on Google Ngram.*

Yet it is true that the terms 'gender dysphoria' and 'gender identity disorder' weakened the inexorability of the link between transgender and sex-reassignment, an outcome more in keeping with the transgender turn than the stricter model of transsexuality.[41] For those critical of any medical models of transgender, however, dysphoria and disorder with their implied mental dysfunctionality and privileging of medical custodianship were still 'slave names'.[42]

Whatever the finer details of the transition and the nature of taxonomic debate, it is remarkable how quickly the term 'transgender' has become established in the American cultural psyche. The Google Ngram provides a rough chart of word usage, with 'transsexual' on an upward trajectory from the 1960s but tapering off from the mid-1990s, whereas 'transgender' increases exponentially from the 1990s onwards, overtaking 'transsexual' at the beginning of the 2000s. See Illustration 22.

Diversity

More to the point, there has been a move from thinking in terms of binaries or dichotomy to accepting ambiguity and diversity. Claudine Griggs's *S/he: Changing Sex and Changing Clothes* (1998) is from that era. However, it is ironic that she focused so resolutely on transsexuality, proclaiming surgery as axiomatic, and denying any 'perceptual middle ground between male and female', when it was written at the very time (Griggs references it) that it was possible to encounter 'transsexuals who do not *want* to alter their genitalia'.[43] Alternatively, Lori Girshick's book, based on interviews with 150 trans-identified people, is subtitled 'Beyond Women and Men', and one of the participants, who described themself as 'ungendered', said, 'I see myself as a gender transcender. I dream of a world where the gender binary is transformed . . . from two genres to an infinite number'.[44] 'How to explain, in a culture frantic for resolution', writes Maggie Nelson, 'that sometimes the shit stays messy?'[45] 'I'm not taking testosterone to change myself into a man', wrote Beatriz Preciado, before he became Paul. 'I take it to foil what society made of me, so that I can fuck, feel a form of pleasure that is postpornographic, add a molecular prostheses to my *low-tech* transgender identity composed of dildos, texts, and moving images.'[46] 'There are not two sexes', he went on to explain, 'but a multiplicity of genetic, hormonal, chromosomal, genital, sexual, and sensual configurations. There is no empirical truth to male or female gender beyond an assemblage of normative cultural fictions.'[47]

An anthology of trans stories had depicted the shift from transsexuality to transgender perfectly, with its narratives of those for whom more radical bodily transformation was not an option and who had never felt trapped inside the body of another sex. Cynthiya BrianKate said, 'I just don't fit the labels and categories most people use for things like gender. I'm not exactly a man or a woman; I'm somewhere in between, a bit of both and neither at the same time.'[48] Phillip Andrew Bernhardt-House described their similar identity as 'metagendered'.[49] The young Preciado had caught the spirit in the late 1990s in their counter-sexual manifesto, republished in 2018, advocating 'somatic communism', the 'systematic destruction of naturalized sexual practices and the gender system', 'a constituting assembly of an endless multiplicity of singular bodies'.[50]

The terms used under transgender recognize more diversity: inter-gendered, omni-gendered, gender fluid, and gender queer, rather than transsexual, MTF, and FTM.[51] Zowie Davy has explained the variety of 'body projects' undergone by the trans men in her 2000s study: 'transmen are phenomenologically too diverse to capture in traditional medical models'.[52] The masculinities of trans men are multiple. Matthew Heinz lists nearly thirty 'identity labels' used by the trans masculine.[53] Some trans men are incredibly attuned to the varieties of being a man. Thomas Page McBee's account of boxing training is a perceptive discussion of New York masculinities in the twenty-first century. 'I was thirty years old and needed to become beautiful to myself', he wrote of his initial injections of testosterone.

> I clocked my becoming primarily in aesthetic terms: the T-shirt that now fit me, the graceful curl of a biceps, the glorious sprinkle of a beard. I loved the way men looked, and smelled, and held themselves. I loved their lank and bulk and ease, their straight-razor barbershop shaves, their chest-first centers of balance. I loved the quiet efficiency of the men's restroom, the ineffable physical joy of running alongside my brother, the shadows we cut against the buildings we passed.[54]

McBee became alert to the burdens of masculinity, implying, when thinking about his own brother, that many men might feel as trapped in their bodies 'as I had in mine'.[55]

Others are more conflicted. 'Lucas felt ambivalent about manhood. Identified as a queer feminist, he struggled with fitting into traditional notions of masculinity that coded as negative (raging, violent, selfish) to him while he also longed for those parts – to be openly angry, aggressive, to fuck like a man.' Lucas was a fictional composite of a number of trans men who have consulted the psychiatrist Griffin Hansbury, trans male himself.[56]

We can see similar diversity in detailed local studies. Twenty-five San Francisco trans men in a 2013 survey varied in their sexual (queer, bisexual, gay, straight) and gender (man, gender queer, trans) identities; all but one were taking testosterone, but many (eight out of the twenty-five) had not had top surgery and only one had had bottom surgery. (Though it should be said that many had rather conventional notions of masculine sexuality.)[57] Twenty-five San Francisco trans women were similarly heterogeneous in their sexual identities (heterosexual, bisexual, gay, lesbian), though unfortunately the researchers did not report their gender identities; all but three were taking oestrogen, most had had cosmetic surgery, but only six out of the twenty-five had had genital surgery.[58]

Given the much longer history of gender and sexual flexibility demonstrated in a previous chapter, the transgender turn might not seem
so momentous. Yet it is undeniable that sexual possibilities multiplied
in the world of postmodern sex. Annie Sprinkle, an articulate presence
at so many moments in the modern history of sex, wrote in 1989
about her first sexual experience with a trans man: 'Having sex with
Les was a constant mind fuck. I could put my finger inside his pussy
... his pussy? ... and feel her balls. His skin was soft and smooth
like a woman's; yet he had hair on his chest.'[59] Bodily transformation
frequently involved what has been termed a new 'sexual habitus', a
new way of being in the social world sexually.[60] 'There has never been
a time that I felt more human, more part of this flesh, than when I am
fucking', wrote one of the trans contributors to a collection of essays
that celebrated this new regime.[61] 'My relationship with my girlfriend
makes me a straight man', a female-to-male transsexual remarked, 'but
I don't feel like a straight man ... I feel politically and culturally like
a lesbian, or actually I'm a man now, I guess I must be gay. But then
I'm not a gay man sexually.'[62] Holly (later Aaron) Devor's work on
female-to-male transsexuals in the late 1980s and early 1990s saw its
point of departure from previous research in terms of its willingness to
recognize sexual indeterminacies and uncertainties and shifting identities: 'FTM TS-identified persons may, at various periods of their lives,
identify as straight, gay, lesbian, men, or women.'[63] Devor was interested in the existence of trans men who identified as gay men, what he
termed 'man-to-man sexuality'.[64] While many of those he interviewed
admitted to a sexual attraction (as men) to homosexual men, few had
actually acted on their desires because of (Devor thought) fear of AIDS
or anxiety about their lack of a penis – perceived to be essential to male
homoeroticism. Nevertheless, one informant admitted to considerable
casual interaction:

> I've actually done it a lot with gay men. And what I do is I go more to
> movie theaters and bathrooms. And I've – it's incredible! Oh, my God.
> It's so wild! It's so amazing! It's such an education! First of all, it's like
> a very intense male bonding thing. I cannot describe it. When straight
> FTMs say to me 'I don't understand how a woman can become a man
> to be gay', I don't understand how they could not at least explore that.

Because it's the ultimate in masculinity . . . How much more masculine can you get? . . . And, of course, it's risky, it's a real adventure.[65]

When Lou Sullivan was trying to transition as a gay trans man in the 1970s and 1980s, he had to form his own networks with other trans men and write his own information booklet. He was a pioneer in not only female-to-male transgender – less commonly recognized in those earlier years than male-to-female – but also as a trans man who wanted to have sex with men rather than women. The gay trans man was barely acknowledged when Sullivan was transitioning; he was refused entry to the Stanford Program twice, primarily because he insisted on his same-sex (male) rather than opposite-sex attractions. Trans men were supposed to desire females in those decades.[66] Yet the gay trans man was no longer such an anomaly by the 1990s. A survey of nineteen Gender Dysphoria clinics that was published in 1995 found that most would 'approve biological females who explicitly stated the wish to live as a gay man postoperatively'. Only two said that they would not approve such patients for sex-reassignment surgery – and one clinic questioned the very existence of such people. While the sixteen positive responses do not guarantee anything other than theoretical endorsement, it is significant that the question was even asked.[67]

In the late 1990s, David Schleifer interviewed five gay trans men, two of whom had lived as lesbians before they had transitioned, and three whose heterosexuality became homosexuality once they had reconstructed their gender. In all cases, their desire for men (as men) was homosexual. Their significance, to quote Schleifer, was that their identities 'presuppose distinctions between males and females, between men and women, and between homosexuals and heterosexuals, while also demonstrating the permeability of these supposedly inviolable distinctions'.[68] These were gay men without penises and with vaginas. And yet they perceived themselves as gay men able to have sex with other gay men: 'I don't have a dick. I don't have a dick in the way most people think of a dick. I actually feel like I do have a dick . . . I really like penetration, in my pussy, in my man-pussy. I don't know what to call it.'[69] The partners of these men were not interviewed but one described a lover as thinking, 'This is fucking a guy who has a pussy; this is something a little bit different.'[70]

'I would describe myself first as a guy, then as a gay man, then as an FtM, then perhaps ... as a queer FtM who still has sex with women (usually butch women or MtFs) once in a while', one trans man observed in a 2009 study.[71] 'I'm a sissy tranny fag. I have a friend who's ftm but not male identified who does female drag. I have third gender friends. It's a GENDER GALAXY.'[72] Women who, as women, had been averse to vaginal penetration enjoyed it as men.[73] Some had had sex with other trans men: 'My first gay sexual experience was at age 26. My partner was another FtM who also identifies as gay.' The partners of the trans men were not interviewed but it was recognized that they added to the sexual complexity: 'it takes an exceptional person to deal with a man who has no penis, but a vagina and a clitoris'.[74]

Already, in the 1990s, researchers were encountering post-operative transsexuals who resisted the obligatory sexual histories:

> I never subscribed to this thing, 'Oh! I'm a woman trapped in a man's body.' I never subscribed to that. I know what I am. I know what I was born. I think that I'm intelligent enough to know that I have never been a woman. I never will be. I cannot profess to know how a woman feels, because I've never been one.[75]

Bornstein said, 'I was never "a woman trapped in a man's body". That is just so much horseshit.'[76] Jason Cromwell thought that, as trans people formulated their own languages for their feelings and experiences, the old categories would be challenged. His insider study – he was trans as well as an anthropologist – stressed that many trans individuals rejected the notion of being trapped in the wrong body: 'It is inconsistent to feel trapped in a category – woman – when one has never felt like others in that category.'[77] Cromwell charted what he termed 'intersections on the transmap'.[78] FTM could stand for female-towards-the-masculine as well as female-to-male. He preferred to think of continuums and of trans people to avoid the fixed identities implied by transsexual termi-nology.[79] His informants included Rich: 'I am one of those who "enjoy my cunt" but still see myself as male. I do not identify as a lesbian or dyke. I am a sexual being and will be sexual with the organs I have.'[80]

A 2004 collection contained reflections by those designated female at birth who found categories limiting.[81] They might take testosterone, but did so 'to become a gender that was neither male nor female'.[82] They rejected the 'correct' way to become trans.[83] They included 'a dude with a vagina' ('What straight man wouldn't want to have a pussy with him at all times?').[84] They were those who had 'transitioned from female to not female'.[85] 'I never chose female and I won't choose male, both are untrue for me.'[86] Several identified as gay males: 'What's a Nice Dyke Like Me Doing Becoming a Gay Man?'[87]

Surveys of transgender people in 2011 discovered hundreds of different self-identity descriptions among trans people.[88] In the study by Genny Beemyn and Susan Rankin of those who classified themselves as 'other' or 'transgender' rather than male or female, the more common identities included genderqueer, androgyne, bigender, 'mostly female', 'I am my own gender', un-gendered, 'both female and male', and 'unsure'.[89] The researchers replaced the traditional categories of male-to-female transsexual and female-to-male transsexual with male-to-female/transgender and female-to-male/transgender and, in recognition of the new sex/gender fluidity, added the descriptors 'male-to-different-gender' and 'female-to-different-gender'.[90] They also discovered flexibility in sexual orientation: those who wrote 'other' instead of gay, lesbian, heterosexual, or bisexual when asked to self-classify.

> In particular, some of the FTM/transgender and female-to-different-gender participants who were attracted to women and who had their roots in the lesbian community continued to identify as 'lesbian', 'queer', 'a dyke', or 'nonstraight' – even though they no longer considered themselves female – because they did not want to erase their pasts and did not think that the label 'heterosexual' adequately reflected their experiences.[91]

Another large study, whose participants were 'Anyone who does not identify strictly as their "male" or "female" birth sex', found that, from a list of provided gender identities, the most commonly chosen, apart from male and female, were genderqueer, transgender, and transsexual.[92] But what was striking was the range of unsolicited descriptions, including varieties of trans as well as 'agender', femme, gender fluid, pangender, and non-gendered.[93] The most favoured sexual orientations were pansexual and queer, but also included bisexual, lesbian, straight, gay, and 'do not identify'. Added descriptions provided by the surveyed included 'both hetero male and lesbian female', homoflexible, 'predominantly heterosexual guy in a girl's body', and 'queerly straight/pansexual'.[94] Significantly, the majority of those surveyed in this online study 'identified as more than one gender identity' and less than half described themselves as transgender.[95] Though it could be argued that, like the sexologists of old, these researchers were instituting as well as capturing categories, given that they provided lists of identities and orientations, their findings do demonstrate, as they claim, trans identity diversity.

Investigations of HIV risk-related activities have generated information on men known as 'trans MSM', transgender men who have sex with (non-transgender) men. These trans men nearly all injected testosterone (though did not share needles) and engaged in potentially high-risk sexual contact, often without using condoms, although their self-reported HIV status prevalence rate was low.[96] They masturbated and had oral, vaginal, and anal sex with men they met on the street or in bars and clubs, and described the attraction of 'sleeping with men who are attracted to men because that's the dynamic'.[97] Their sexual partners may or may not have known that their encounters were transgender: 'I hook up, and so I'll very often tie a guy up and not disclose. I kind of jerk them off or something like that. And so I think sometimes they think its odd that I'm not like whipping out my dick and jerking off or something, but mostly they're too busy. They're having a good time.'[98]

In terms of fluidity of gender identity and sexual orientation, it is notable that these trans men claimed multiple identifications: male, trans, gender queer; queer, gay, bisexual, and even (though only for one man) straight.[99]

Gender and sexual flexibility seem to be the big winners in these accounts. The classic manifestoes of queer studies and queer theory of the 1990s had had almost nothing to say about trans matters (the most that they managed was a few pages on 'drag' and 'cross-dressing').[100] But by the end of the first decade of the new century, queerness was ubiquitous in the trans community. Bay Area trans men in a 2015 study – who identified variously as 'man', 'trans man', 'trans boy', and 'genderqueer' – mostly described their sexuality as 'queer'.[101]

Drag Kings

One of the more interesting products of the transgender turn is the phenomenon of the drag kings, women who perform masculinity.[102] In the classic *The Drag King Book* (1999), Jack Halberstam distinguished between the 'butch' drag king who performs 'what comes naturally' and the 'femme' or androgynous drag king for whom the performance of masculinity was 'an act'.[103] The former performed a version of her/his own masculinity; the latter abandoned her masculinity when she left the venue. Some of the early studies of drag kings stress their lesbianism.[104] But drag kinging was in transition as Halberstam was writing, for, as he acknowledged, many drag kings could be better described as trans or gender queer. Some of the facial hair was real.[105] Halberstam's co-author, the photographer Del LaGrace Volcano, saw the drag king as 'part of the transgendered spectrum'.[106] A study of one group, the Santa Barbara-based Disposable Boy Toys, active from 2000 to 2004, found that half of them saw themselves as 'genderqueer', 'transgender' or 'FTM', rather than as simply female.[107]

Essentially, the bodies of the drag kings proclaim a female masculinity, or rather masculinities – 'rescuing masculinity from the clutches of maleness' as Kathryn Rosenfeld has articulated it.[108] Both Volcano's photography in *The Drag King Book* and Gabriel Baur's moving images in the documentary feature *Venus Boyz* (2001) demonstrate a multiplicity of being male: from bare-chested jean-wearing to suited man-about-town, from the white London working-class geezer to the black New York homeboy, and from hip hop coolness to lounge bar sleaze. It shows a costumery of flannel shirts, baggy jeans, basketball shirts, boxer shorts, leathers, and puffer jackets; a range of facial hair (goatee, beards, designer stubble, moustache, sideburns), chest hair, hair style (tapered, buzz cut, grunge, comb over, pompadour, crew, slicked back, shaven, Afro, quiff, side part), and male-themed accoutrements (braces, ties, belts, boots, shoes, bandanas, hats, caps, and . . . dildos).[109]

Some kings see their masculinity as corresponding to a better sense of their identity. Johnny Science, a trans man, who founded The Drag King Club in New York in the early 1990s, gravitated into the gay male S/M scene. He also established The Jovian Gentlemen, a gay male social club.[110] He had earlier initiated the FTM Fraternity, a support group for trans men 'in all stages of transition . . . What brings them together is the challenge of living their lives as men.'[111] New York kings had firm links to the trans community. A Cleveland drag king, Xavier Alexander Jade, claimed that he 'found out what "trans" was through drag'.[112] Svar Tomcat said that his performance persona was not unlike his non-stage personality: 'my masculinity may be more emphatic, but it is not essentially different'.[113] Silas Flipper pointed out that male clothing was not really drag for him: 'if I were to wear drag it would be female clothing'.[114] For MilDred Gerestant (Dred), on the other hand, the main appeal of drag kinging is the expression of gender fluidity.[115] The same applies for the veteran cross-dresser Peggy Shaw. 'I have been described as masculine', she wrote in 2010; 'Actually I am a new kind of femininity. I am interested in testing masculine–feminine and butch–femme as markers. I want to go way beyond the boundaries of the girls' room and the boys' room.'[116] Other drag kings adopt masculinity to reprise aspects of maleness. Club Geezer in London mocked British masculinities – some of the images from those performances are very amusing.[117] One of the (affectionate) targets of Mo B. Dick (Maureen Fischer) is the 'typical Brooklyn guy who mouths off, "I ain't no homo" and "suck my dick" and "fuck you" . . . I see this as total parody and I get off on emulating maleness in such an extreme and crass way.'[118] However, for Dred and Shon, two black drag kings who perform contemporary R&B, their act is a homage born of respect rather than satire.[119] But, whatever the motive, this playing with gender destabilizes binaries of masculine and feminine.[120] One of the Disposable Boy Toys said that drag was 'the gateway drug for gender regardless of what that gender is'.[121]

We have seen that there were male impersonators in the past, and WOW, the 1980s New York women's theatre collective, put on plays with cross dressed women playing male roles.[122] However, it remains true that drag *kings* were a product of the 1990s, able to be located fairly precisely in New York in 1990 in workshops run by Diane Torr and the trans man Johnny Science.[123] Science produced and fronted a public access television programme, *The Drag King Club*.[124] He is pictured, bearded, with a group of drag kings who appeared on *The Joan Rivers Show*.[125] See Illustration 23. The ubiquitous Annie Sprinkle was an early drag king. There is a Polaroid of her as a biker in the same collection in

23 *Drag kings on* The Joan Rivers Show, c.1995.

the Fales Library.[126] In fact, New York at that time was a vibrant scene of female gender bending: a flyer for a social group called AmBoyz refers to butches, gender benders, drag kings, boychicks, she-bears, tomboys, amazons, female guys, boss girls, and sirs, as well as trans men and transsexuals.[127] So the kings were archetypically of the transgender turn.

Drag Queens

The drag queens have a much longer heritage. We have already discussed the earlier period, but male cross-dressing was very much part of the transgender turn, too, burgeoning in the queer nightlife of the cities, among the performers and participants in club, bar, and house/ball culture (as we will discuss later). Nan Goldin's photography captured the sheer beauty of the New York drag queens in the early 1990s, in the bars and clubs and working as models for *Vogue*. See Illustration 24. Some were transsexual; some were gay cross-dressers; others, she wrote, 'live in a gender-free zone, flaunting their third sex status'.[128] It is a rather ephemeral world, appearing fleetingly in zines like New York's *Pansy Beat* of 1989–90, for example, or in labours of love such as Julian Fleisher's illustrated field guide to the same city's drag scene in the 1990s.[129] Amos Badertscher's extensive photographic portfolio of the clubs and bars in Baltimore and Washington, DC shows the sheer inventiveness and vivacity of the male femininities proclaimed in those arenas.[130] The fierce, Latina(o) Sista Face, hostess at Washington's Tracks and Velvet Nation in the 1990s, asserted a formidable gender hybridity. See Illustration 25. Badertscher wrote that her 'drag magic' made 'club nightlife more entertaining, exotic, exclusive'.[131] Or there are the stunning, gender-blending, androgynous images of the club aficionado, John Flowers, who was a patron/performer (it is impossible to distinguish between the two) at the same DC clubs. See Illustration 26. Tracks was, Badertscher recalled, 'the most original, inventive, technologically stunning and outrageous' gay club of the 1980s and 1990s, 'the best show in town! Any race imaginable and any sexual persuasion just had to show up there.'[132] The striking, though very different, femininities of Sista Face and Flowers were declared as much in the absence as in the presence of clothing, which makes them all the more intriguing.

24   *Nan Goldin, Jimmy Paulette, and Tabboo! in the bathroom, NYC 1991.*

Verta Taylor and Leila Rupp's study of gay-identified drag queens at
the 801 Cabaret in Key West in Florida from 1998 to 2001 is the best
recent academic study of the genre. Some of the performers interviewed
for the study contemplated 'sex-reassignment surgery'; one referred to
being a 'closeted transgendered person'.[133] Another said that she felt
like herself when in drag.[134] But others wanted to perform femininity
rather than to achieve it permanently. As Sushi, one of the performers,
proclaimed: 'A drag queen is somebody who knows he has a dick and two
balls.'[135] In retrospect, Taylor and Rupp thought that the drag queens
had 'very little knowledge of transgender'.[136] Texas drag queens studied
by a linguist in the 1990s 'were careful to distinguish themselves from
transgender women'.[137] But the researcher Rusty Barrett thinks that
drag has shifted since then, away from realism towards overstatement,
because of the impact of transgender. 'Thus drag has come to draw on
more exaggerated looks that make it clear that the performer is indeed
a drag queen and not a (cisgender or transgender) woman.'[138]

Some of the Key West drag queens were obviously men dressed as
women; others were far more gender-ambiguous in their presentation. However, all – in either their verbal repartee, jokes, song, dress,

25   *Amos Badertscher, Sista Face, 1997.*

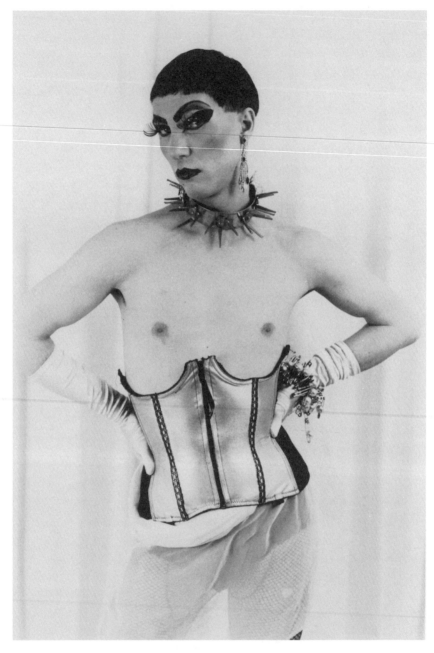

26   *Amos Badertscher, John Flowers, 1999.*

interaction with the audience, or visual performance – challenged notions of gender and sex. Dean described his drag persona, Milla, as 'omnisexual': 'I, Milla, is a black woman, she is a Puerto Rican woman, she is a lesbian, she is a heterosexual woman, she is a gay man.'[139] Another joked that she was bisexual: 'Buy me something, I'll be sexual.'[140] Kylie used the lines that she is getting moist in her non-existent vagina, and that the model Naomi Campbell has a gorgeous body just like hers apart from the dick.[141] One of the most interesting aspects of Taylor and Rupp's study is that they interviewed audiences as well as the performers, and found that those watching the drag queens felt that their gender and sexual categories were challenged in the space and moment of performance.[142] As one drag queen challenged an audience member in a Philadelphia-based study, 'I can be your girlfriend . . . I will let you suck my cock.'[143]

Drag has had a heavy theoretical burden to bear, especially in the time of transgender. For Carole-Anne Taylor, writing in 1991, the themes of gay male drag were misogyny, phallic narcissism, masochism, and, of course, disidentification. 'The gay man in drag, like other men in a patriarchal symbolic, may feel whole at woman's expense, since he too can refuse her difference, misrecognizing it as lack and fetishistically disavowing even that.'[144] Judith Butler used drag in her famous rethinking of gender as performance: 'In imitating gender, drag implicitly reveals the imitative structure of gender itself – as well as its contingency.'[145] She went on to dispute the stability of any original of which drag was a parody. Drag threatened the original itself. As she put it later, revisiting that earlier work, the example of drag was meant to 'make us question the means by which reality is made'.[146] And for José Esteban Muñoz, drag, as performed by Vaginal Davis in the 1990s, becomes disidentificatory (Davis: 'I wasn't really trying to alter myself to look like a real woman'), intended to critique, unsettle categories, 'to create critical uneasiness and, furthermore', in the words of Muñoz, 'to create desire within uneasiness'.[147]

Decolonizing Transgender

We should be wary of privileging the 'transgender model'. The favouring of the US experience has distorted transgender analysis, warping indigenous constellations of gender.[148] At the beginning of this book, we touched on the nation's own two-spirit people, who were in categories of their own – flexible, changing, combining elements of male and female, but of neither.[149] A Lakota informant told Walter Williams in 1982 that they 'would be terribly scared to be a man or a woman . . . *Winktes* are two spirits, man and woman, combined into one spirit.'[150] While we do not need to go as far as Will Roscoe in proclaiming that America might once have been 'the queerest continent on the planet', it is certainly significant that the continent contained the multiplicity of genders that Americans in the twenty-first century can only dream of.[151]

But I have in mind more global examples of the colonizing gesture. There are many to work with, but a few should give a sense of my thinking. The Brazilian *travestis* are homosexual men with breasts and a penis, who construct themselves as female to attract men but who (despite their name) are neither transvestites nor transsexuals. They do not want to be women. As Don Kulick has observed, they are unusual in that they 'irrevocably alter their body to approximate that of the opposite sex without claiming the subjectivity of that sex'.[152] Similarly, the Mexican *vestidas*, whom Annick Prieur encountered in the late 1980s and early 1990s, wore women's dress and make-up, grew their hair long, tucked their genitals, and enhanced their breasts, hips, thighs, and buttocks, either through padding or by injecting oil, silicone, or hormones. They sculpted female bodies but were homosexual men or *jotas*, rather than transsexuals. They are, wrote Prieur, 'recognized as non-males inside the category of males'.[153] Or there are the 'multifarious forms of gender/sexual variance' in India and South Asia, the '*kothi–hijra* spectrum' of multiple and shifting identities so brilliantly discussed by Aniruddha Dutta and Raina Roy, impossible to contain under the limiting designation of transgender.[154] Can we decolonize in the sense of reversing the process: how can the varied apprehensions of gender and sex in other cultures – which we should resist calling transgender – help us to rethink the received pattern in the USA? Can they challenge those very hegemonic assumptions?

One of the strengths of what has come to be called intersectionality is that it takes seriously the dynamics of race and class so vital to the history of trans.[155] Race was there in the 1950s in the whiteness of the first celebrity trans, Christine Jorgensen, as Emily Skidmore has outlined.[156] The founders of the early 1970s trans activist group STAR had long complained that gay liberation was a 'white, middle-class, white [*sic*] club'.[157] Struggling to survive on the streets as black and Latina cross-dressed sex workers, dealing with rape and trying to avoid death ('I think I'm like a cat. I've almost lost my life five times'), they had almost nothing in common with their fellow liberationists.[158] 'Gay liberation but transgender nothing!' as one of those founders, Sylvia Rivera, conveyed it from the perspective of 2001.[159] Race is evident too (along with the privileges of wealth) in the most recent example of white celebrity trans, Caitlyn Jenner, a proclaimed role model who seems so out of step with everyday trans lives. The crafted trans self-presentations on YouTube, too, are predominantly white moving images.[160] Heinz has observed of the Internet that 'The dominant image of a transman is that of a white, middle-class, educated, able-bodied, young man, either baring his chest or presenting in a *Gentleman's Quarterly* pose.'[161] Where are the representations that black trans men and women can identify with? Hence, Kortney Ryan Ziegler's documentary *Still Black* (2008), and his statement, nearly ten years later (2017), that trans men were still visually marginal, were prompted by the invisibility of black trans men in media culture.[162]

Yet trans people of colour are central to the transgender turn in the USA. Jeff Cowen's prints from the 1980s and Katsu Naito's comparable photographs for the 1990s portray black and Latina(o), New York, trans street culture.[163] See Illustration 27. There is a vibrancy and dignity in these images of trans people living lives of adversity. They seized the initiative in posing: 'street models', is how the Introduction to Naito's book described them.[164] Or they were photographed merely going about their daily or nightly business. Susanna Aikin's documentary *The Salt Mines* (1990) was, we will see, about homeless Latina(o) trans sex workers.[165] Daniel Peddle's *The Aggressives* (2005) dealt with Puerto Rican and black female masculinity.[166] Taking race and class seriously provides a different perspective. As Jack Halberstam has written, examples of gender variance in people of colour mark 'the whiteness of the category "transgender"' and suggest 'that other terms exist in other communities and that these other terms indicate the function of gender in relation to a specific set of life experiences'.[167] Certainly, the

27   *Katsu Naito, 'West Side Rendezvous, 1992'.*

'masculine of center', trans men of colour in a recent collection called *Trans/Portraits* (2015) insist that their identity is tied up with their race and sexuality as much as their gender, and feel alienated by the associated whiteness of transsexuality/transgender.[168]

One study of female-identified trans communities in New York in the first decade of this century argued that ethno-cultural links were more important than any common sense of trans-ness. What did Black and Latina(o), working-class, house/ball aficionados of Harlem, the Bronx, Queens, and Brooklyn have in common with Asian sex worker 'transvestites' and girls of Manhattan and Queens, or with white, middle-class, cross-dressers in Manhattan?[169]

Studies of house/ball culture in Newark, New York, Detroit, and elsewhere have located blended queer and trans-related categories of identity, even within these rather confined Black and Latina(o) communities.[170] The butch queens (BQ) are gay males, assigned male at birth, but who can present as masculine or effeminate. 'Their presentations range from "straight acting" to "flamboyant".'[171] The femme queens (FQs) are trans women – in various stages of transition. The butch queens in drag (BQs up in drags) are gay males who perform in drag. The butches are trans men or cross-dressed women, or masculine lesbians.[172] In fact, Marlon Bailey refers to a 'six-part gender system' in ballroom culture, also including women (assigned female at birth) identifying variously as lesbian, queer, or straight, and 'men/trade' (assigned male at birth) and straight-identified – though presumably sexually available for queer sex.[173] Sex, gender, and sexuality, writes Bailey of ballroom culture, 'are malleable and mutable'.[174] The 'fluidity of both gender and sexuality makes identification in Ballroom complicated and multidimensional'.[175] For Lucas Hilderbrand, the appeal and power of the documentary *Paris Is Burning* (1990/1), centred on the New York City drag balls of the 1980s, was its 'testimony that there are other ways of being in the world, ways that are self-defined'.[176] Gerard Gaskin, who has produced an insider's photographic archive of the balls from the 1990s to the 2000s (in New York, Philadelphia, Richmond, and Washington, DC), refers to a conscious gender and sexual fluidity, where 'house and ball members perform what they wish these cities could be'.[177]

'I'm tired of being labeled', Sylvia Rivera wrote in the early 2000s, looking back on a hard life of drug use, sex work, and trans activism, 'I don't even like the label *transgender* . . . I am Sylvia Rivera. Ray Rivera left home at the age of 10 to become Sylvia. And that's who I am.'[178] It is important to note that many of the participants in the trans-sexual moment and the transgender turn did not even use the words transgender or transsexual. Ballroom culture thought of femme queens and butches rather than transgender.[179] Hilderbrand has mused that the protagonists in *Paris Is Burning* (focused on 1987) probably did not think in terms of transgender.[180] Brooke Xtravaganza had had implants and surgery – what she called 'a transsexualism operation': 'I'm no longer a man. I am a woman.'[181] Venus Xtravaganza wanted to 'have a sex change' to 'be a complete woman'.[182] Octavia St Laurent hoped, by 1988, 'to be a full-fledged woman of the United States'.[183] But Pepper Labeija said, 'I've been a man, and I've been a man who emulated a woman. I've never been a woman . . . I never wanted to have a sex change'; 'I'm gay.'[184] True, Angie Xtravaganza talked of 'being a transsexual in New York City'.[185] Yet, generally, the elements of transgender that were there were so in the absence of the actual terminology. The master category in *Paris Is Burning* was gay: 'gay life', 'being gay', 'the gay world'.[186]

An early 1990s study of trans sex workers in Atlanta found that most identified as gay – though, as with many in *Paris Is Burning*, they called each other 'she', used female names, and wore female dress.[187] The male-to-female sex workers whom David Valentine talked to in New York in the late 1990s, 'the Meat Market girls', saw themselves as women and gay; they rarely acknowledged the term 'transgender'.[188] Hence, Fiona said both that she had always been gay (which assumed sexual attraction to men as a man) and that she had 'been a woman' all her life.[189] Anita said 'I know I am gay and I know I'm a man' but 'I treat myself like a woman . . . I do everything like a woman'. She knew that she was not transsexual but had never heard the term transgender.[190] Mona likewise was gay in her attraction (as a male) to other men, but was also a woman. Valentine observed that these subjects did not adhere to the strict categories that had originally informed his research agenda: 'for both Anita and Mona, "liv[ing] as a woman" does not preclude being "gay" where "gay" indexes erotic desire for someone who is male-bodied'.[191] (He discovered that those who came into most contact with the various social service agencies were more likely to begin to use the more exclusive category of transsexual and to distinguish it from homosexuality.[192]) Miss Angel, another of his contacts, was 'simultaneously gay, homosexual, transsexual, and "a woman with a large clit"'.[193] Valentine found a blending of categories – 'transgender, gay, transsexual, transvestite, and others' – in a milieu where he had expected the terms 'transsexual' and 'transgender' to impose more uniformity of self-identity.[194]

Interestingly, Valentine's findings reprise the world of the Latina trans women, crack cocaine users who had featured over ten years earlier in Aikin's *Salt Mines* (1990), living in abandoned trucks in the Department of Sanitation's salt storage area by the Hudson, and eking out an existence ('We survive') through sex work in New York's Meatpacking area, dangerous streets, the site of the earlier-mentioned photography of Cowen – now, incidentally, the location of high-end fashion stores, restaurants, and the (new) Whitney.[195] See Illustration 28. Cowen, who took a tape recorder as well as a camera when he talked to these women, said that some merely dressed as women because they could make more money that way, and some 'felt more comfortable being women'. But the 'large majority have not not had sex changes because of lack of funds, but mostly because they view themselves as being neither male nor female, but, being something of both. So they get their breasts enlarged, but are still proud of their penises.'[196] Aikin was unsure of how to describe these individuals when she encountered them: 'They were men dressed as women. But not just men dressed as women, they were – I guess the word is transvestites, but that is not what entered my mind. At that moment I was just struck by their hybrid beauty. They looked like feminized men or masculine women. Something in between both sexes.'[197] In the film, we see them maintaining their femininity: washing, depilating, taking hormone shots, obtained under the counter without prescription ('Here are the hormones'), and putting on make-up.[198]

Some of these women used the language of transgender. 'Since I can remember', said Gigi, 'my feelings are those of a woman. As they say, I am a woman encased in the body of a man.' 'Why would I dress as a male if I don't feel like one?', she replied to a rather inept question from the filmmaker.[199] Giovanna observed that she was a 'transvestite': 'I accept it . . . I'm a transvestite . . . to the fullest'.[200] Yet they more frequently described themselves as 'drag queens' or queens. Giovanna was a 'drag queen' as well as a transvestite and stressed her attachment to both her penis and her breast implants. Her sexual attraction to men was expressed in homosexual terms.[201] '"We're a family of Latin

28 *Jeff Cowen, Greenwich Street, 1988.*

queens", drawled Gigi.'[202] Sara referred to herself as a 'gay person' and even Gigi, when describing her attachment to her boyfriend Edwin, communicated it in terms of love 'between two men'.[203]

Despite questions about being 'trapped in the wrong body' and 'being a preop transsexual', the female-identified, Southern, 'black gay men' interviewed in 2004 and 2005 by E. Patrick Johnson varied in their perceptions of who they were, although their church affiliations complicated the picture considerably.[204] One, who had lived as a woman for many years, working as a hair stylist and performing on stage in drag, dressed as a man to attend church. She had contemplated surgery, but decided that, having transitioned in every way but the physical, surgery seemed somewhat superfluous.[205] Another, a gay preacher who performed drag, was uncertain of his transness. Although he had been cross-dressing since he was a teenager, he was adamant that 'We didn't do it at that time because we wanted to be women or I was a woman trapped in a man's body. I didn't feel that, you know, not at all.'[206] Yet another identified as a gay man who was a transvestite, a gay man who loved women's clothing: 'I identify with gay men. I identify with cross-dressers as well. I'm not interested in changing my sex.'[207]

More recently, there are other examples of those included under the trans umbrella who do not self-classify as such: 21st-century, New York, house/ball goers, African American, Black, and Latina(o) trans women, many of whom engaged in sex work, described themselves as '*fem queens, nu women*, or *girls*' rather than as transvestites, transsexuals, or transgender.[208] Studs, females of colour, who celebrate their masculinity and date femme women but who do not aspire to actually be men ('I don't wanna be no man'; 'I ain't no tranny'), either distinguish themselves from transgender and transsexuality, or do not much use the terms at all.[209] One, Desiree, said that she had not heard of transgender before she attended a LGBTQ centre. Having been introduced to the term, she thought it might describe her situation, but the point is that she had existed quite well without it, seeing herself as a stud and lesbian. They did not suffer in the absence of trans labels. They were still able to identify their sense of self. D'Andre, who, when asked, said that she was 'technically transgender' but 'not transsexual' (so was familiar with the terms), said, 'I can't call myself a man and I don't feel like a woman, so I just identify as stud . . . It's a mix between the two.'[210] Paloma had

'only vague knowledge' of transgender terms, and told those who interviewed her that she wouldn't use transgender because 'Then they're
gonna start saying, "She got testicles and stuff."'[211] The Los Angeles
stud Lolo asked her therapist, Vernon Rosario, about the possibility
of hormones to modify the size of her breasts and to masculinise her
voice but 'did not feel she needed to pass for male, nor did she insist
that people think of her as male or refer to her with male pronouns'.[212]

Then there are the Aggressives, masculine-presented Puerto Rican,
Asian, and black women who hover between transgender and lesbi-
anism, the subject of Peddle's documentary *The Aggressives* (2005).
Marquise worked on his masculinity: his body, jawline, wearing a sports
bra and bandaging his breasts. He competed in the ball scene with
Aggressives who were taking hormones – and 'they ain't got no titties'.
Marquise had contemplated taking hormones – 'guys grow up' – but
Aniche, his girlfriend, was not keen on it: 'A breast with facial hair, it's
wrong.' Marquise said that he considered himself 'to be transgendered,
but, no, I don't want to be a man, I like lesbian women', explaining that
he was not one of those who considered themselves to be a straight
man who will only date straight women. 'I'm a lesbian, I'm just very
aggressive.' He presented as masculine in everyday life, but when with
his girlfriend they are lesbians: 'I eat her muffin, she eats mine.'[213] 'I
consider myself a faggot', Tiffany tried to explain; 'I have a femme
aggressive attitude, I guess you could say. I'm not butch, I'm just me.'
(His rap act certainly included basic male misogyny.) 'I'm attracted to
a female appearance, that's why I don't identify as a lesbian, cos I'm
[he was with Kelly, a trans woman at that time] not a lesbian if I date
transgenders and have heterosexual sex with them. I don't date men,
like, you know.'[214] Rjai wears men's suits and is very masculine-looking.
He has won a lot of awards at the balls. 'I am a woman, so what if I'm
gay.' 'It was easier looking like a male.' 'I feel like a woman every day.
My sexuality has nothing to do with my gender. I have male-like ways, I
have female-like ways.'[215] Again and again, they resisted categories: 'I'm
bein' me' (Octavia); 'I'm just me' (Kisha).[216]

The Oakland-based Brown Boi Project has used the term 'masculine
of center' as a term that includes masculine-oriented people of colour,
including the Aggressives and studs.[217] The category of transbutch
attempts to combine trans masculinity with a sense of being a woman.

Much like the studs, Aggressives, and brown bois, 'Transbutches embody a . . . space . . . between the uncritical celebration of "womanhood" and its rejection entirely'.[218]

Non-trans men who have sex with transgender women (known as MSTW or, more vernacularly, as 'admirers') also threaten categories. The Johns Hopkins researchers John Money and Margaret Lamacz once invented a technical name for them, gynemimetophiliacs, and gynemimetics for the object of their desire, but the terms did not catch on.[219] In a sense, MSTW are part of the trans world but are not trans themselves, though they have trans-directed desires. (Consumers of such pornography – derogatorily, known as she/male porn – are expressing such desires in what has been called the '"transsexualization" of the heterosexual male'.)[220] MSTW are not 'homosexual' and usually say as much, though their sexual partners may categorize them as closeted – and, less politely, as 'just another faggot who likes to take dick up his ass'.[221] The drag queen International Chrysis reputedly said that such men 'just want to feel that big dick up their ass while those titties are on their back. 'Cause in their mind they're having sex with a woman, but its homosexual sex, which is what they want! They want gay sex.'[222] If pressed, as they have been in some studies, MSTW may self-classify as 'bisexual' or 'homosexual'.[223] But they are not bisexual in the sense of being attracted to either male or female. As one man put it, quoted by David Valentine, he was only attracted to 'women and transwomen'.[224] Those MSTW who say that they are heterosexual are clearly not conventionally straight: 'despite the openness with which Mark frankly voiced his enjoyment of a shemale penis, the discordance between the penis as an object of desire and his view of himself as a heterosexual man presented Mark with a dilemma'.[225] As a trans character in a Jonathan Ames novel told a protagonist (who, like the book's author, was attracted to trans women), 'You're a typical trannychaser. You're not really straight, but you're not really gay. You're straightish.'[226] A New York MSTW who liked receiving anal sex from trans women with big penises described himself as 'heterosexual with a twist'.[227]

Although Martin Weinberg and Colin Williams divided their San Francisco interviewees into heterosexual and bisexual, they actually concluded that identity was not the best way of understanding these men. 'It is our contention that the main focus of those with a sexual interest in transwomen was not in self-identity but in a sexual release fostered by a tolerant environment [bar culture] that allowed for the pursuit of easy sex.'[228] Another San Francisco study encountered men who did not subscribe to any orientation: '[I am] just sexual. I don't have any orientation'; 'I'm a try-sexual. I'll try anything.'[229] Or, as an admirer told Daniel Mauk for his ethnography of New York MSTW,

> Admirers are just attracted to something different and something special. You don't have to be gay to go with a transgender woman. You don't even have to be gay to go with a person of the same sex, in my opinion. What is bisexual, gay, straight? There's a very thin line between these [categories] . . . for the admirers, in my opinion, it's just a matter of sexual desire.[230]

Where do non-trans women who are attracted to trans men sit in the spectrum? Transamorous and transsensual are sometimes used to describe them; 'tranny chaser' is more frequent but offensive.[231] As one such woman explained, she had identified variously as lesbian, queer, and pansexual, but now did not identify at all:

> I don't subscribe to labels right now because I don't think I'm straight, and I don't think I'm gay, and I don't think I'm bi 'cause I'm not really attracted to women. I know I'm not attracted to women. And I'm not attracted to cisgendered men. But when you tell people that you're mostly attracted to trans guys, then they call you a tranny chaser.[232]

So, claims for the greater inclusivity of 'gender identities' under the trans umbrella may be somewhat beside the point.[233] Should transgender studies 'dispense with identity as an analytic trope'?[234]

Surgery Again

It is inconceivable that the transgender turn was not influenced by a reaction against some of the traumas of gender affirmation. State-of-the-art summaries of transsexual surgery contained imagery of forearms marked out for the fleshly flaps that comprised the phallic shaft and the urethra, and the less-than-convincing products of that surgery. They showed the removal of the penile skin, the division of the urethra, litigated stumps, the creation of a vaginal cavity, the turning inside out of the penile skin tube, clitoral release, the extension of the urethra, and labial manipulation: all in graphic detail.[235]

Claudine Griggs's marathon – hour-by-hour – account of her vaginoplasty in Dr Stanley Biber's famous clinic in Trinidad in Colorado in the early 1990s is unforgettable in its gruesome chronicling of the pain of what has been called the 'gothic narratives of bloody dilations of her new vagina'.[236] It is an account all the more powerful because it is so beautifully written, has so much obsessive detail, and for the fact that we know that it had not originally been intended for publication: 'I wanted a detailed chronicle that years hence would recall the sex-change experience without the distorted fondness I sometimes project into distant memories, and the pain in body and mind screamed for non-distortion.'[237] 'About 2:00 P.M.', she writes in a chapter called 'Pain', 'Pain; more pain; I am having a baby, ripping me apart.' 'About 6:30 P.M.', she continues on the same page, 'I'm nauseated and vomit fiercely. I don't mind the nausea itself; the convulsive agony is so overwhelming that nausea compares as a minor discomfort.'[238]

Suzanne Kessler's *Lessons from the Intersexed* (1998) used 1990s transsexual Internet discussion groups – containing in-house information – as a compelling source for the shortcomings of the surgery performed on both transsexuals and intersexuals: 'You refer to your results as "semi-ok", I call mine bad.'[239] With admirable understatement, Kessler observed that the 1990s was 'not an era in which genital surgery has been perfected'.[240] Perhaps it is significant, in retrospect, that Anne Bolin's study of trans women, which so emphasized the role of surgery in transsexuality, did not actually have a chapter on surgery (there was one on hormonal management). The details of this pivotal process were discussed in a matter of a few pages, though it was admitted that 'The operation is a major trauma to the body.'[241]

'I didn't know there was so much pain in the world', was the 2003 entry in *The Diary of a Transsexual Academic*; 'I felt like I had been split in two by the experience.'[242] Recalling earlier surgery, April Ashley wrote in 2006 of pain 'the most acute a human can experience', 'a most hideous pain' – though she did not regret the operation.[243] The shock of surgery in Juliet Jacques's 2012 online accounts of blood and pus and pain, and hyperventilation at the sight of the wounds, shows that the drastic interventions have continued.[244] In the Afterword to its second edition in 2004, Griggs said, unsurprisingly, that she knew of people who had been put off surgery because of her book.[245] Even *Hung Jury* (2012), intended as a counter to the negativity surrounding genital surgery, and brutally honest in its life stories, has enough descriptions of agony, excessive medical costs, protracted complications, and disappointments, to daunt the most determined trans man.[246] 'Since 2005 I have had seven major surgeries performed in hospitals and at least 16 more procedures performed in the surgeon's office', wrote one of the contributors to *Hung Jury*. 'I will have spent in excess of $100,000. And no, I still won't be done . . . Phalloplasty should not be taken lightly.'[247] Many trans people must have sought alternative, more benign, adjustments.

Surgery was not limited to the genitals, of course. Facial feminization surgery (FFS) has become increasingly favoured in the last twenty to thirty years, especially among more affluent trans women (its lack of appeal for trans men is interesting).[248] But its bone grinding, cutting, fracturing and resetting, its soft-tissue and cartilage slicing and reshaping, was not without its pain – as brows were reduced, noses reshaped, cheek bones enhanced, chins shortened and sharpened, jaws reduced, scalps advanced, hairlines reshaped, eyebrows raised, upper lips shortened, lips augmented, and thyroid cartilage tissue removed.[249] We should acknowledge, but probably not set too much store in, the accounts of disgruntled patients, given that the endorsements of FFS seem to be so glowing.[250] But if any prospective patients read Eric Plemons's brilliant account of the 'visceral materiality' of his day observing in the operating theatre ('Is that brain or sinus?'), it should surely give them pause.[251]

For many of the accounts in *Hung Jury*, the challenges of genital surgery represent the forging of masculinity: 'like being reborn from the fire. I am emerging stronger than I could have imagined.'[252] The FFS patients whom Plemons talked to describe their experience as 'climactic', 'cathartic', (in his words) 'the actualization of this idealized possibility'.[253] So the avoidance of surgery has probably had more to do with the costs involved in transsexual transition. '[P]ay for it yourself or go without', is CN Lester's pithy description of trans medical care.[254] Most people cannot afford gender reconstruction. Similarly with facial feminization surgery, not usually covered by medical insurance, the costs of up to $60,000 can be prohibitive.[255] Plemons describes FFS as 'a luxury available only to the resourced few'.[256] As Caitlyn Jenner put it when she decided to have FFS and breast augmentation, 'I also have the luxury of being able to afford such an extensive procedure.'[257]

Molly Haskell wrote her memoir of her new sister's transition in the 2000s in order to advance transgender understanding, but it also underlines the sheer cost of that journey: the facial reconstruction (forehead, nose, chin, tracheal shaving), electrolysis, the clothes, wigs, deportment and voice training – quite apart from the genital modification. 'It was just so important for me to get it all as perfect as possible', her sister said of her face work. 'I feel bad for those transsexuals, so many of them, who can't afford it, because the first thing you look at is the face. I was fortunate enough to have enough savings to do it.'[258] But contrast this wealthy, aged, established, white experience with that of the young, black, poor, trans woman Janet Mock, padding her hips with stolen shoulder pads, saving her lunch money to buy a friend's hormone tablets, gaining access to a sympathetic endocrinologist and what she called an 'underground railroad of resources', and then engaging in sex work (Mock's *Redefining Realness* provides an insider's account of 'survival sex work') as she moved through school and university and towards gender reassignment in Thailand at the age of 18.[259]

As we saw of an earlier era, there remains a tendency to self-medicate because of the problems of access and cost. CN Lester claims from personal knowledge that 'self-medication is extremely common'.[260] Toni Newman, who worked 14th Street in New York in the 1990s, recalled another black trans sex worker, known as 'The Doctor', who sold silicone shots.[261] Pumpers inject silicone into the calves, hips, buttocks, and face, to provide cheaper (though not inexpensive) feminization – often with disastrous health implications. A report in the *New York Times* in 2011 claimed that the practice was widespread among the immigrant and poor population – it was estimated that 22 per cent of trans women in New York had been pumped.[262]

Complexities

I have been underlining the gender and sexual flexibility of transgender
against the rigidity of the transsexual binary – what Katrina Roen has
characterized as 'both/neither [man or woman] versus either/or [man
or woman]'.[263] Yet Roen discovered that the trans people she talked to
in the late 1990s actually negotiated between these 'apparently competing discourses'.[264] Similarly Sally Hines's work on trans identities in
the UK demonstrates that the so-called umbrella includes both the
'fluidly situated and queerly practised' (transgender) and the 'more
consistently formed' (transsexual) – and that people might shift from
one to the other, depending on context, life stage, and audience, when
articulating their own identities.[265] Some of those who self-identified
as transsexual to Erin Calhoun Davis in 1999 and 2000 said that it
enabled them to acknowledge a 'blend' of genders and that their past
history was part of their identity.[266] Zowie Davy has been persuasively
critical of pitting gender queerness against transsexuality, arguing for a
greater recognition of the gender multiplicity in individual transsexual
stories.[267] Juliet Jacques's is one such story. She grew up in the 1990s,
was reading trans history and theory in the 2000s, including the work of
Bornstein, was well attuned to the complexities of transgender, and was
critical of simplistic notions of being trapped in the wrong body. Yet
she still opted for transsexual genital surgery.[268] The transgender turn
was more complicated than we might imagine.

   J. R. Latham has detected transsexual traces in transgender medicine: the persistence of certain assumptions about being trapped in the
wrong body, adjusting the body to match the mind, and the singular
path of rectifying that perceived incongruence.[269] A singular transsexuality is axiomatic – '*assumed* in very narrow and restrictive terms' – that
neglects, disavows, the intricacies of trans lives.[270] As Kadji Amin has
expressed it, stereotypical autobiographical models of transsexuality,
while empowering the subjectivities of some, have invalidated those of
others.[271]

   Latham's brilliant account of navigating the Australian transgender medical gatekeepers – doctors, psychiatrists, and surgeons – is a
telling example, by a highly articulate and knowledgeable observer/
participant, of all the limitations of the transgender turn.[272] As he

negotiated his breast surgery in 2010, he continually encountered the lingering stereotypes of former regimes: the wrong-body narrative; a demonstration of what might be termed genital alienation; a focus on 'sex as genital', rather out of keeping with modern trans sensibilities.[273] In fact, his claims are strikingly similar to the findings of a study of US medical attitudes in the 2000s which detected entrenched binary views of sex and gender and a privileging of heteronormativity straight out of 1960s transsexuality.[274]

Latham felt, just as he might have felt had he been going through the same process decades earlier, that certain narratives, longings, and expectations had to be articulated to convince those who determined the eventual surgery: 'Sex is a convincing narrative.'[275] Latham disconcerted his doctors, psychiatrist, and surgeons because he refused to take testosterone before surgery (the usual ordering), was reluctant to take hormones at all, was not alienated from his female genitalia, and did not hate being female. As he summarizes his predicament, 'I am being compelled to articulate and enact sex (and transexuality [*sic*]) in a specific way, a way that is limiting my ability to articulate my actual experiences and desires.'[276] The gatekeepers had their own assumptions about the road to masculinity and the qualities that demonstrated their patient's bona fides – attitudes not vastly different from those of Harry Benjamin's transsexual handbook of 1969.[277]

At much the same time, in San Francisco, Nick (then Nina) Krieger went through a similar process before their top surgery, acquiring a letter of support from his therapist, 'my piece of paper, the key that so many had been forced to acquire by reiterating a standard narrative, the myth of a sole transgender experience that bore no resemblance to mine'.[278] When he published his account of his experiences, Krieger made sure to include a counter-narrative: 'born female, Nina resembled a girl, then woman, then boy, and is on her way to becoming, well, I have no idea. But whatever the ending, it will be happy.'[279]

Rhyannon Styles's recent autobiography is interesting for its demonstration of both the promises and confines of the transgender turn. She was conscious of the role that gender-confirming surgery played in her transition: 'I had journeyed through my transition regarding a neo-vagina as the finishing line.'[280] She recalled that in 2012 she did not really consider that she had any choice in her quest for feminizing treatment: 'I thought that I had to say that I wanted a vaginoplasty in order for my prescriptions to be given to me.'[281] But when the surgical moment came in 2015, she was more aware of gender flexibility, of those who identified as 'non-binary', of the possibility of being a woman with a penis ('I have never hated my penis').[282] So she did not have the surgery and describes herself as 'trans-femme'.[283] It was a transgender choice, but it could so easily have been a transsexual one.

The wrong-body narrative still arguably dominates popular conceptions of trans bodies.[284] Looking back on her life as Bruce, Caitlyn Jenner wrote: 'The only thing missing is that I have this woman living inside of me.'[285] Michael Lovelock has argued that this discourse has meshed with a wider culture that more generally (especially for female bodies) sees 'the transformation of the physical body ... as a means of releasing or affirming one's essential subjectivity'.[286] As the trans woman Laverne Cox said in *TRANSform Me*, a 2010 reality show in which trans experts gave makeover advice to non-trans women, 'We're all transgender women, and because of that journey ... to bring out the woman that we've always felt that we were inside, we are uniquely qualified to help you do the same thing.'[287] Transsexual essentialism flourishes.

So, too, does the lingering requirement to desexualize trans experiences when attempting to tell life stories and/or to access the required care. Latham has noted that any complexities, and, to be sure, sexuality itself, are ignored in many autobiographies and in the narratives required by the medical experts – the two are linked, of course.[288] Jacqueline Rose was also 'struck at how little space for sex' there was in the transsexual narratives that she read in preparation for her article on trans that was published in the *London Review of Books* in 2016.[289] And this is in the face of the sexual and gender flexibility that, for whatever reason (choice or necessity), was self-evidently at the heart of the sexuality outlined earlier in this chapter. Sexual expression is actually pivotal to the transgender turn and to the 'myriad ways' that trans people make sense of their gender.[290] Indeed, a recent collection has reinstituted the *sexual* in transsexual as a way of demonstrating 'the effervescing and unpredictable force of sexuality'.[291] As Bailey has said of the Detroit ballroom, 'the gender system ... is always about sexuality'.[292] Or, as the trans man Paul B. Preciado wrote in 2018, 'the libidinal transformation is as important as the epistemological one: desire must be transformed. We must learn how to desire sexual freedom.'[293]

Children

It would be inaccurate to say that the trans child is a product of the transgender turn. We saw (in Chapter 4) that children played a role in the transsexual moment; Karl Bryant, one of those children, went on to produce a very good Sociology Ph.D. on that topic.[294] Julian Gill-Peterson has written a whole book countering the 'myth that there were no trans children until recently'.[295] However, the child has certainly featured far more in recent trans history.[296]

During the transsexual moment, as we saw, the aim was to dissuade children from a transsexual path, but the belief was that if this had not been achieved at a very young age – by 5 or 6 – then the therapist could cautiously explore 'the patient's wish to live as a member of the opposite sex', as Lawrence Newman from the Gender Identity Clinic at UCLA put it in 1970.[297] With the transgender turn, however, facilitation can occur from the outset: the trans or gender-nonconforming children in Ann Travers's study *The Trans Generation* are as young as 4.[298] Those therapists who wish (as Newman would have articulated it) to 'reverse' or 'prevent' cross-gender identification are increasingly out of step with current trans care norms – hence the closure of the Child, Youth and Family Gender Identity Clinic in Toronto in 2015.[299] As a matter of fact, its Head, Kenneth Zucker, formed a direct link with gender identity research in the 1970s. His book, published in the 1990s as a state-of-the-art summary of gender conflict in children, looked back to Richard Green's early work on trans children, and discussed similar rationales for treatment: reduction of social ostracism, treatment of underlying pathology, prevention of adult transsexuality, and avoidance of homosexuality.[300] A more recent discussion in 2008 had Zucker still advocating interventions to reconcile children to the sex that they were assigned at birth, with techniques highly reminiscent of 1960s and 1970s therapy, and with undue attention to the behaviour of the parents ('parental psychopathology').[301]

Zucker's approach contrasts with what is now called the gender affirmative model, where, in reversal of the 'adult–child dynamic . . . the child becomes the source of knowledge, albeit temporarily and only in certain moments'.[302] Children's apprehensions of their gender identity are treated seriously, as early as 18 months to 3 years: 'We define *gender health* as a child's opportunity to live in the gender that feels most real or comfortable to that child and to express that gender with freedom from restriction, aspersion, or rejection.'[303] As one of the mothers in Tey Meadow's study of trans kids told her child Hunter, a 6-year-old trans boy, 'Sure honey ... lots of boys have vaginas.'[304] Rather than dissuading children from their gender creativity, it is facilitated, and puberty is delayed by 'puberty blockers' to allow greater time for decision-making and to facilitate any future transition by means of hormones or surgery.[305] The goal is developing gender, rather than reversing or correcting it.[306] The older approach is increasingly seen as 'the pathology approach', whereas the more recent strategy is 'the affirmative approach'.[307]

'If you want to get really, really specific', one of Travers's trans kids told her, 'I am a demi-polyromantic, polysexual, gender-queer individual'.[308] 'Some days I'm masculine, and that's pretty weird', Cameron told Susan Kuklin for her book on transgender teens. 'Some days I'm feminine, and that's pretty weird too.'[309] The gender and sexual identities of trans children, especially teens, reflect the flexibility discussed at the beginning of this chapter. Travers was struck by what they describe as the 'living dictionary of identity and practice' encountered in the current trans, gender-nonconforming generation. 'The way that gender- and sexual-minority people of all ages produce language beyond the limits of binary gender, heterosexuality, and monogamy is an exciting thing to chronicle.'[310] Or, as Meadow has voiced it, 'It is rare to have an opportunity to watch an emergent social category in formation. Transgender children provide us with precisely this opportunity.'[311]

Conclusion

Finally, what of the proclaimed transgender tipping point, invoked in this book's Introduction? In her study of queer youth at a drop-in centre, Mary Robertson was aware that this protective hub of gender experimentation and flexibility – and the promise of a nonbinary future – existed in a wider American society that, despite an increased awareness of transgender, was much less forgiving when it came to ambiguity.[312] A New Museum of Art collaboration, *Trap Door*, has warned against becoming mesmerized by a claimed cultural/visual acceptance at a very time when it is patently dangerous to be a trans person, especially a trans woman of colour.[313] The 'trap door' in the collection's title indicates a period of what Jeannine Tang has called 'heightened vulnerability' alongside the 'rise in visibility'.[314] Rogers Brubaker has warned perceptively that, while the 'shift toward public acceptance of transgender has been astonishingly rapid, . . . it has been uneven across regions, generations, institutions, and milieux'.[315] In fact, a cluster of studies of trans rights attests to that unevenness, precariousness, and provisionality.[316]

It has been a troubled period of recent history, with all the hyperbole about 'trans moments' masking continued prejudice, discrimination, and violence. It can include what has been called 'mundane transphobia', a gender normativity assuming the right to ask the most ill-informed, inappropriate, personal questions, illustrated fully in Oprah Winfrey's interview with the pregnant trans man Thomas Beatie: 'Did you like have a penis implant or you did something else?'[317] Increased conspicuousness brings a greater likelihood of what Christopher Shelley has termed 'repudiation', which goes way beyond mere transphobia. 'In their status as Others', he writes, 'transpeople are frequently positioned within an alterity that is not of their own making; whether they like it or not, they are often forced to endure the primal reactions of others.'[318] If this was true in 2008, how much more applicable is it ten years later? Even the well-intentioned can be condemnatory; 'if you think about it a little longer', writes the trans man Maxim Februari, 'you will see that compliments about my bravery are in fact signs of disapproval'.[319]

Gender-rigid prisons are especially harsh towards transgender women, with their personnel as implicated in the violence as other prisoners.[320] 'While in prison, I have been cut by other prisoners for refusing to perform sexual acts for them, and I have been beaten and sexually assaulted (sodomized with the nightstick) by correctional staff.'[321] Rape is axiomatic. Studies of California prisons have found that some 60 per cent of transgender women inmates have reported sexual assault, and many engage in protective sexual pairings to avoid violence from other prisoners.[322]

And then there is the violence done to identity in a system that is incapable of accommodating anything other than gender binaries or of considering gender beyond that assigned at birth.[323] For the reality of these gender-rigid regimes is that their occupants exhibit obvious, though unacknowledged, gender fluidity.[324] A study of trans women in prisons in Pennsylvania found an inability, shared by fellow prisoners and officials, to think of them as anything other than homosexuals (because they were seen as men rather than women).[325] The trans men or gender-nonconforming in women's prisons are forced into an alien femininity. Julia C. Oparah has put it nicely: not all those in women's prisons are women, and not all incarcerated women are in prisons for women.[326]

'Transgender Murders: 2017 Off to a Horrific Start', the trans woman Lexie Cannes blogged at the beginning of that year.[327] The same could be said of every year. Cannes's site, *State of Trans*, certainly posts a depressing chronicle of brutality and hatred towards trans women.[328] Transgender Europe's Trans Murder Monitoring figures make for horrifying reading, with nearly 3,000 tabulated deaths worldwide in the ten years from 2008 to 2018, and with over 200 in the USA alone. Trans people, many of whom are sex workers, are being shot, stabbed, beaten, and strangled to death. The statistics show that the 'street' is by far the most common location of death.[329] The *Remembering Our Dead* website provides a complementary, highly poignant, register of trans deaths and suicides as they happen, linking them to newspaper accounts and other Internet sites, memorial cards, photographs, and Tweets from TransLivesMatter.[330] The Facebook group Transgender Violence performs a similar role.[331] Of course, this particular strategy of memorialization may obfuscate the racial and class context of this carnage and the role of the state in its perpetuation – Sarah Lamble has written a persuasive critique[332] – yet these chronicles of death and social violence still puncture any societal self-congratulation about trans moments. As one of the contributors to *Trans/Portraits* said in relation to LGBT celebrations, 'We're still too plagued by death to be able to celebrate pride.'[333]

Conclusion

In 2017, in the aftermath of Donald Trump's victory, a cohort of trans-masculine, trans-feminine, and gender-nonconforming people from New York, San Francisco, and Atlanta told Walter Bockting and his fellow researchers that they remained resilient but feared for their future. Theirs was an accurate appraisal of the vicissitudes of trans history. Greater trans visibility, the so-called tipping point, brought with it the possibility (especially in the light of the 2016 election) of reaction: a backlash. One of the survey group, a young trans man from New York, referred to a 'two steps forward, one step back problem'; and the researchers invoked Martin Luther King about a long road to justice.[1] Yet Bockting and his colleagues could have turned to trans history itself. As we have seen in this book, an earlier period of trans visibility in the 1960s and early 1970s faced a vehement backlash right at the time that trans, in the form of what was termed transvestism and transsexuality, seemed to be so ascendant.

It is a complex trans past. I have written elsewhere about the gay trans man Lou Sullivan, but he is important for this book, too.[2] See Illustration 29. His diaries record an oscillation between feelings of transvestism and of transsexuality; it was certainly no easy narrative of a man trapped inside the body of a woman, the plot that trans men were encouraged to adopt during that period. Sullivan worried that the

29   *Lou Sullivan, 1988.*

construction of a penis would affect the ability to orgasm. He vacillated between discomfort with ambiguity and anxiety about giving in to what he thought, and feared, was fantasy.

The story of Sullivan is compelling because it threatens the very categorization that his biographical sketch posthumously forced him into. His diaries and correspondence with other trans people reveal that he moved in a world of gender and sexual indecision and flexibility. Lou knew a trans man who had lived with a trans woman for thirteen years, when, for much of that time, 'she was he and I [the trans man] was she. We had heterosexual sex for 7 years.' The man was interested in having sex with women but was also attracted to 'Male to females . . . Women with penises'. Lou was puzzled by this irresolution:

> You stated 'I try to think what it would be like to be gay as I never had sex with a genetic female.' But . . . if you feel like a man and have only had sex with men, wouldn't that make you a gay man?? I'm confused. You love women but have never had sex with them? Is it that you feel like a man, and just WANT to have sex with a female that makes you heterosexual??? Help me understand what you mean.[3]

Another friend oscillated between maleness and femaleness and the names that proclaimed their identity. She sought and then decided against surgery (admittedly when it was refused by Cook County Hospital), stopped taking hormones, went back to them again, and then stopped. She threw out her women's clothes. She cut her hair short. 'I'm really back? (forward!) to being MALE.' Lou had a photograph of her on a Florida beach, an image of an attractive, bare-breasted, tanned woman. She and Sullivan discussed their inability to fit categories. 'We're further [a]head than any of those TV–TSs. To be able to incorporate your ambiguity is far more revolutionary & futuristic than trying to resolve it along some obscure conventional lines.' They talked of living 'in the middle', being neither male nor female.[4]

Lou corresponded with a lesbian who was attracted to trans men, with a trans man who had sex with bisexual men but who also 'turned tricks' as a 'gay male', and with a trans man who had had another trans man as a lover – 'Do you use your vagina during sex? (I mean during male/male sex)'.[5] He heard of a woman who had once been attracted to gay men (her husband had a male lover) but then decided that she was probably lesbian.[6] In response to a personal advertisement in the magazine *Coming Up* (1984–5), he was contacted by a trans woman who was 'strongly lesbian identified' and liked to be sexually dominated: Lou replied, 'we really are coming from opposite directions'.[7]

Those with whom Lou/Sheila had sex, as both/either Lou/Sheila, are as interesting in their flexibility as our main protagonist. In 1980, when he was undergoing hormone treatment, Sullivan went to a bar and picked up a drag queen. Sullivan paid for their drinks: 'She sat down & I was the man, she was the loose woman.' They went back to Lou's place for sex. The drag queen said to Lou that he looked a little effeminate (she told him later that she had thought that he was intersex). The drag queen had anal sex with Lou and told him that he was 'going to make a really great gay boy'![8] The diaries, then, record a woman, Sheila, who wanted to be, and then decided that he was, a man – a homosexual man. Sheila/Lou enjoyed sex with men, but as a gay man. As a famous punk band member told his occasional girlfriend Sheila, 'she' had indeed set up a challenge for 'herself'.[9]

Sullivan's diaries provide a sense of the fluid sexual identities in San Francisco in the late 1970s and the 1980s as the transsexual moment turned into transgender. The then Sheila went to a drag bar and groped and kissed a 'black queen'. A young man joined them and then became agitated when he discovered that the queen was male. He became even more disturbed when he was told that Lou was female: he had thought that the female was male and the male female. Sullivan writes in his diary that it would not have made any difference: it would have been a threesome. 'God, if he'd 've done a 3-way with a girl & a guy, why not after we turned out to be a girl & a guy anyway?'[10] By 1981, having had both hormone treatment and a mastectomy, Lou had decided that he was transsexual rather than transvestite, writing to Virginia Prince, the well-known male-to-female transvestite, 'Maybe I never was a true TV. I never wanted to be a female, even a female dressed in men's clothing. I just wanted to be a gay man.'[11] Meanwhile, he participated in homosexual sex as a homosexual man, avoiding contact that would reveal any femaleness. He engaged in oral and masturbatory sex, even visiting gay porn theatres. Some men discovered Lou's female bodily attributes, but did not care; others did not seem to notice. The casual encounters could pass without too much interrogation – 'I just cannot believe all the fucking *action* I'm getting in the fucking Mission District.'[12] Longer-term relationships proved more awkward: 'But what ARE you? You're not a man & you're not a woman! . . . You have a pussy but you're still a man.'[13] Sullivan had a heterosexual boyfriend who had sex with women and vaginal sex with Lou, but tired of the arrangement: 'I really am bored with this heterosex we have constantly.'[14] He also had sex with a male-to-female transvestite; 'she says I'm the best man she's ever had . . . and she's had REAL ones. Too bad I can't get too excited about her.'[15]

We have seen trans in its various forms – what Rogers Brubaker has termed the trans of migration (transsexuality), the trans of between (gender blending), and the trans of beyond (those who transcend categories, though, ironically, they find a new category as nonbinary).[16] We have examined the time before transsexuality, when those aware of their nonconforming gender managed an existence – even sought out surgical and hormonal solutions – under a cultural framework where inversion (as it was known) was interpreted as a species of homosexuality. We have encountered people who lived their lives almost oblivious to the medical model, or who consciously carved out ways of being and seeing that did not adhere to the dominant categories of gender and sexuality or of transsexuality and transgender. We have met both those adept at negotiating their way past the gatekeepers and those who avoided them completely: self-help is a recurring theme in trans history. Trans was formed in the streets as much as it was in the consulting room and surgery. We have noted the importance of cross-dressing in trans before trans, the transsexual moment, and the transgender turn – a neglected, vital strand in both American and trans history. We have located the importance of trans people of colour, despite the dominance of whiteness in much trans imagery. We have discovered that the sexual and gender flexibility viewed as so central to the transgender turn had earlier precursors. We have established the importance of sexuality in the history of trans. Despite the eagerness of some trans observers to separate gender from sexuality, bodily variety was accompanied by diversity of desire. We have charted constantly shifting concepts of trans from a time before it was named to its current visibility, identities blurred and challenged to the point where trans itself seems on the point of inaugurating a new 'nonbinary turn'.

Above all, we have seen that it is important to recognize transgender as process. The Canadian, gender-nonconforming performer Rae Spoon has outlined how they have lived such a gender and sexual history, first as a woman attracted to other women (lesbian), then as a trans man dating women (heterosexual), then as a man with male partners (gay) – sometimes cis men, sometimes trans men – and then as gender-neutral, using the personal pronoun 'they', what they refer to as retiring from the gender binary.[17] But they also want to 'leave room for future possibilities' that they 'have not been presented with yet'.[18] It is an apt summary of trans history.

I began this book with a criticism of the historical amnesia of some trans studies, so it is only fair to give credit where credit is due. Juliet Jacques, who was a teenager in England in the 1990s, tried to make sense of her identity, almost oblivious to the era in which she lived: 'I'm gay and I wear women's clothes . . . I thought that made me a drag queen.'[19] 'Unknown to me as a teenager, there was an explosion of theory in North America during the decade as authors explored their identities through a mixture of autobiographical, historical and feminist writing.'[20] But she soon made up for it, devouring works of trans history and theory. She came to know the words 'transvestite' and 'transsexual', but did not think that they described her situation, and the concept of being trapped in the wrong body 'never quite spoke to me'.[21] She appears to have started using the word 'transgender' slightly before she was introduced to the work of Kate Bornstein in 2004, ten years after it was first published. Jacques read culture intelligently, searching not exactly for role models but for those who might hold a clue to her own gender identity, including some names in this book: Andy Warhol's cross-dressed superstars, Jayne County's memoir, Jack Smith's *Flaming Creatures*. She first encountered Candy Darling at remove, played by Stephen Dorff in *I Shot Andy Warhol* (1996), but later, when she hunted down Morrissey's films from the earlier period, she saw the transgender star herself on screen in *Women in Revolt* (1971). Jacques drew on this cultural history (much of it American) to make sense of her experience.

Or there is CN Lester, who takes history seriously in their transgender education. They have written movingly of the role of knowledge of the past in their own journey.

> What would it mean, to trans people now, if our history were common knowledge? By this I do not mean a history of people who are exactly as we are, regardless of the dictates of historical and social context, but of the people whose lives and efforts helped to create the categories, the structures and cultures through which we now move.[22]

Just a fragment of such knowledge, Lester continues, would have helped them understand themself, providing 'the comfort of companionship, when I was invisible and alone'.[23] 'What a tremendous gift it would have been, to have known that there were people in that history who might now be called trans, people who lived in the genders they *knew* that they were, regardless of what society had told them.'[24] Perhaps some readers will use my book in a similar way.

While I am critical of any notion of easy progress towards some current, historical high point, it does not hurt to finish (like Bockting and his fellow researchers) with a touch of optimism. At the end of the second decade of the twenty-first century, I have students – a few – for whom gender flexibility is the new normal, and who, when it comes to sex, consider themselves queer with a straight twist rather than 'mostly straight'.[25] I have others – more – who are sympathetic to such views. Surely that is some reason for hope for the future.

Notes

Introduction

1 M. L. Brown and C. A. Rounsley, *True Selves: Understanding Transsexualism – For Families, Friends, Coworkers, and Helping Professionals* (San Francisco, 1996), p. 25; J. F. Boylan, *She's Not There: A Life in Two Genders* (New York, 2003), p. 174.

2 R. Erickson, 'Foreword', in R. Green and J. Money (eds.), *Transsexualism and Sex Reassignment* (Baltimore, 1969), p. xi.

3 M. W. Valerio, *The Testosterone Files: My Hormonal and Social Transformation from Female to Male* (Berkeley, 2006), p. 2.

4 E. B. Towle and L. M. Morgan, 'Romancing the Transgender Native: Rethinking the Use of the "Third Gender" Concept', *GLQ*, 8:4 (2002), 469–97. *True Selves* comes in for specific criticism in this regard (at 478).

5 C. Millot, *Horsexe: Essays on Transsexuality*, translated by K. Hylton (New York, 1990), p. 141. First published in French in 1983. Robert Stoller termed transsexualism 'a newly described condition (the literature begins only in 1953)': R. J. Stoller, *The Transsexual Experiment* (London, 1975), p. 2.

6 J. Meyerowitz, *How Sex Changed: A History of Transsexuality in the United States* (Cambridge, Mass., 2002), p. 1. See chs. 1 and 2.

7 H. Benjamin, *The Transsexual Phenomenon* (New York, 1966);

Green and Money (eds.), *Transsexualism and Sex Reassignment*; Stoller, *Transsexual Experiment*. The term was really first used in print by the popular sexologist D. O. Cauldwell in 1949, as will be discussed in chapter 2.

8 The best history of transsexuality is Meyerowitz, *How Sex Changed*.

9 Ibid., pp. 217–22.

10 For an excellent short history of US transgender, see G. Beemyn, 'US History', in L. Erickson-Schroth (ed.), *Trans Bodies, Trans Selves: A Resource for the Transgender Community* (New York, 2014), ch. 22. The essay was published separately in a longer version as an Ebook: G. Beemyn, *Transgender History in the United States* (New York, 2014).

11 See, for example, R. Ekins and D. King, *The Transgender Phenomenon* (London, 2006); S. Stryker, P. Currah, and L. J. Moore, 'Introduction: Trans-, Trans, or Transgender?' *Women's Studies Quarterly*, 36:3–4 (2008), 11–22; S. Stryker, *Transgender History* (Berkeley, 2008, 2017).

12 A. Bolin, *In Search of Eve: Transsexual Rites of Passage* (New York, 1988), p. 84.

13 W. O. Bockting, 'Psychotherapy and the Real-Life Experience: From Gender Dichotomy to Gender Diversity', *Sexologies*, 17:4 (2008), 211–24, quote at 214.

14 H. L. Talley, 'Facial Feminization and the Theory of Facial Sex Difference: The Medical Transformation of Elective Intervention to Necessary Repair', in J. A. Fisher (ed.), *Gender and the Science of Difference: Cultural Politics of Contemporary Science and Medicine* (New Brunswick, NJ, 2011), ch. 10; E. Plemons, *The Look of a Woman: Facial Feminization Surgery and the Aims of Trans-Medicine* (Durham, NC, 2017), quote at p. 1.

15 D. Denny, 'Interview with Anne Bolin, Ph.D.', *Chrysalis Quarterly*, 1:6 (1993), 15–20; A. Bolin, 'Transcending and Transgendering: Male-to-Female Transsexuals, Dichotomy and Diversity', in G. Herdt (ed.), *Third Sex, Third Gender: Beyond Sexual Dimorphism in Culture and History* (New York, 1996), ch. 10.

16 G. Beemyn and S. Rankin, *The Lives of Transgender People* (New York, 2011), pp. 23–6.

17 J. W. Wright, *Trans/Portraits: Voices from Transgender Communities* (Hanover, NH, 2015).

18 Ibid., pp. 72–3.

19 L. M. Diamond, S. T. Pardo, and M. R. Butterworth, 'Transgender Experience and Identity', in S. J. Schwartz and others (eds.), *Handbook of Identity Theory and Research* (New York, 2011), ch. 26, quote at p. 630.

20 The title of the journal's double inaugural issue, 'Postposttranssexual: Key Concepts for a Twenty-First-Century Transgender Studies', *TSQ*, 1:1–2 (2014).

21 For trans*, see S. Stryker and P. Currah, 'Introduction', *TSQ*, 1:1–2 (2014), 1–18, quote at 3. See also A. Tompkins, 'Asterisk', *TSQ*, 1:1–2 (2014), 26–7.

22 Tompkins, 'Asterisk', 27.

23 A. Z. Aizura, *Mobile Subjects: Transnational Imaginaries of Gender Reassignment* (Durham, NC, 2018), pp. 11–12.

24 M. Rajunov and S. Duane (eds.), *Nonbinary: Memoirs of Gender and Identity* (New York, 2019).

25 CN Lester, *Trans Like Me: A Journey for All of Us* (London, 2017), Ebook, loc. 521.

26 The editors, 'Future Gender', *Aperture*, 229 (2017), 23.

27 E. J. Green, *Young New York* (New York, 2019).

28 J. Rose, 'Who Do You Think You Are?' *London Review of Books*, 5 May 2016.

29 For *Transparent*, see S. Stryker and others, 'Virtual Roundtable on *Transparent*', *Public Books*, 1 August 2015: www.publicbooks.org/ artmedia/virtual-roundtable-on-transparent.

30 For television and trans children, see A. Prochuk, 'From the Monster to the Kid Next Door: Transgender Children, Cisgender Parents, and the Management of Difference on TV', *Atlantis*, 36:2 (2014), 36–48, quote at 37.

31 For example, J. A. Peters, *Luna a Novel* (New York, 2004); M. Ewert, *10,000 Dresses* (New York, 2008); B. Katcher, *Almost Perfect* (New York, 2009); J. Carr, *Be Who You ARE!* (Bloomington, Ind., 2010); C. Beam, *I Am J* (New York, 2011); C. Kilovadis, *My Princess Boy* (New York, 2011); K. Cronn-Mills, *Beautiful Music for Ugly Children* (Woodbury, Minn., 2012); K. E. Clark, *Freak Boy*

(New York, 2013); A. Fabrikant, *When Kayla Was Kyle* (Lakewood, Calif., 2013); A. Gino, *George* (New York, 2015); D. Gephart, *Lily and Dunkin* (New York, 2016); M. Russo, *If I Was Your Girl* (New York, 2016). I am grateful to Claire Gooder for compiling this list.

32 For Jenner, see the two-season reality series, *I Am Cait* (2015, 2016); *Vanity Fair: Trans America*, Special Edition, 18 August 2015; and C. Jenner and B. Bissinger, *The Secrets of My Life* (London, 2017). For Mock, see J. Mock, *Redefining Realness* (New York, 2014); J. Mock, *Surpassing Certainty* (New York, 2017); and her blog https://janetmock.com.

33 M. Lovelock, 'Call Me Caitlyn: Making and Making Over the "Authentic" Transgender Body in Anglo-American Popular Culture', *Journal of Gender Studies*, 26:6 (2017), 675–87.

34 L. Horak, 'Trans on YouTube: Intimacy, Visibility, Temporality', *TSQ*, 1:4 (2014), 572–85; T. Raun, 'Archiving the Wonders of Testosterone Via YouTube', *TSQ*, 2:4 (2015), 701–9, quote at 701; T. Raun, *Out Online: Trans Self-Representation and Community Building on YouTube* (London, 2016). See, too, M. Heinz, *Entering Masculinity: The Inevitability of Discourse* (Chicago, 2016), which discusses trans masculinity on YouTube.

35 D. Udy, '"Am I Gonna Become Famous When I Get My Boobs Done?": Surgery and Celebrity in *Gigi Gorgeous: This Is Everything*', *TSQ*, 5:2 (2018), 275–80.

36 Erickson-Schroth (ed.), *Trans Bodies, Trans Selves*.

37 C. S. Salgado, S. J. Monstrey, and M. L. Djordjevic (eds.), *Gender Affirmation: Medical & Surgical Perspectives* (New York, 2017).

38 W. O. Bockting and J. M. Goldberg (eds.), *Guidelines for Transgender Care* (Binghamton, NY, 2006); A. E. Eyler, 'Primary Medical Care of the Gender-Variant Patient', in R. Ettner, S. Monstrey, and A. E. Eyler (eds.), *Principles of Transgender Medicine and Surgery* (New York, 2014), ch. 2, quote at p. 26.

39 E. Coleman and others, 'Standards of Care for the Health of Transsexual, Transgender, and Gender-Nonconforming People, Version 7', *International Journal of Transgenderism*, 13 (2011), 165–232, quote at 168; G. Knudson and others, 'Identity Recognition Statement of the World Professional Association for Transgender

Health (WPATH)', *International Journal of Transgenderism*, 19:3 (2018), 355–6, quote at 356.

40 S. Stryker and S. Whittle (eds.), *The Transgender Studies Reader* (New York, 2006); S. Stryker and A. Z. Aizura (eds.), *The Transgender Studies Reader 2* (New York, 2013).

41 A. Haefele-Thomas, *Introduction to Transgender Studies* (New York, 2019).

42 E. Shipley, 'Etymology', in T. C. Tolbert and T. T. Peterson (eds.), *Troubling the Line: Trans and Genderqueer Poetry and Poetics* (Callicoon, NY, 2013), pp. 193–4, quote at p. 193.

43 A. H. Devor, *The Transgender Archives: Foundations for the Future* (Vancouver, 2014); A. H. Devor and L. Wilson, 'Putting Trans* History on the Shelves: The Transgender Archives at the University of Victoria, Canada', in A. L. Stone and J. Cantrell (eds.), *Out of the Closet, Into the Archives: Researching Sexual Histories* (Albany, NY, 2015), ch. 10.

44 K. J. Rawson, 'Transgender Worldmaking in Cyberspace: Historical Activism on the Internet', *QED: A Journal in GLBTQ Worldmaking*, 1:2 (2014), 38–60, quote at 38; www.digitaltransgen derarchive.net/about/overview.

45 www.lib.umn.edu/tretter/transgender-oral-history-project.

46 Z. Drucker and R. Ernst, *Relationship* (New York, 2016); M. Seliger, *On Christopher Street: Transgender Stories* (New York, 2016). The quote comes from Mock's 'Foreword' to Seliger, p. 12.

47 https://broadlygenderphotos.vice.com.

48 https://originalplumbing.bigcartel.com/category/magazines. See the recent sample: A. Mac and R. Kayiatos (eds.), *OP Original Plumbing: The Best Ten Years of Trans Male Culture* (New York, 2019).

49 R. Bastanmehr, 'Is Pop Culture Having a Trans Moment?' *Vice*, 3 November 2014: www.vice.com/read/were-having-a-trans-moment-456.

50 K. Steinmetz, 'The Transgender Tipping Point', *Time*, 29 May 2014; K. Steinmetz, 'Beyond He or She: How a New Generation is Redefining the Meaning of Gender', *Time*, 27 March 2017.

51 *National Geographic*, January 2017.

52 M. C. Burke, 'Resisting Pathology: GID and the Contested

Terrain of Diagnosis in the Transgender Rights Movement', *Advances in Medical Sociology*, 12 (2011), 183–210.

53 For early statements, see M. D. O'Hartigan and R. A. Wilchins, 'The GID Controversy', *Transgender Tapestry*, 79 (1977), 30–1, 44–5.

54 For a perceptive discussion of these tensions, as they relate to both medicine and law, see J. L. Koenig, 'Distributive Consequences of the Medical Model', *Harvard Civil Rights – Civil Liberties Law Review*, 46 (2011), 619–45.

55 R. Lane, '"We Are Here to Help": Who Opens the Gate for Surgeries?', *TSQ*, 5:2 (2018), 207–27, quote at 208.

56 A. H. Johnson, 'Normative Accountability: How the Medical Model Influences Transgender Identities and Experiences', *Sociology Compass*, 9:9 (2015), 803–13.

57 M. J. Hird, 'A Typical Gender Identity Conference? Some Disturbing Reports from the Therapeutic Front Lines', *Feminism & Psychology*, 13:2 (2003), 181–99, quote at 183.

58 Z. Davy, 'The DSM-5 and the Politics of Diagnosing Transpeople', *Archives of Sexual Behavior*, 44:5 (2015), 1165–76; G. Davis, J. M. Dewey, and E. L. Murphy, 'Giving Sex: Deconstructing Intersex and Trans Medicalization Practices', *Gender & Society*, 30:3 (2016), 490–514.

59 D. Spade, 'Mutilating Gender' [2000], in Stryker and Whittle (eds.), *Transgender Studies Reader*, ch. 23, quote at p. 326.

60 J. Butler, 'Doing Justice to Someone: Sex Reassignment and Allegories of Transsexuality', *GLQ*, 7:4 (2001), 621–36, quote at 632.

61 M. Violet, *Yes, You Are Trans Enough: My Transition from Self-Loathing to Self-Love* (London, 2018), Ebook, locs. 2636–40.

62 L. M. Lothstein, 'Group Therapy with Gender-Dysphoric Patients', *American Journal of Psychotherapy*, 33:1 (1979), 67–81, quotes at 71.

63 Ibid., 75.

64 A. C. Keller, S. E. Althof, and L. M. Lothstein, 'Group Therapy with Gender-Identity Patients – A Four Year Study', *American Journal of Psychotherapy*, 36:2 (1982), 223–8, quote at 224.

65 E. J. Windsor, 'Golden Ticket Therapy: Stigma Management

Among Trans Men', in O. Gozlan (ed.), *Current Critical Debates in the Field of Transsexual Studies* (New York, 2018), ch. 9, quotes at pp. 134, 135.

66 N. Krieger, 'Writing Trans', in Erickson-Schroth (ed.), *Trans Bodies, Trans Selves*, pp. 582–3, quote at p. 583.

67 See E. Plemons and C. Straayer (eds.), *TSQ*, 5:2 (2018): *The Surgery Issue*.

68 Plemons, *Look of a Woman*, p. 17.

69 M. Davidson, 'Seeking Refuge Under the Umbrella: Inclusion, Exclusion, and Organizing Within the Category *Transgender*', *Sexuality Research & Social Policy*, 4:4 (2007), 60–80.

70 Ibid., 66.

71 R. Connell, 'Transsexual Women and Feminist Thought: Toward New Understanding and New Politics', *Signs*, 37:4 (2012), 857–81.

72 K. Bornstein and Z. Drucker, 'Gender is a Playground', *Aperture*, 229 (2017), 24–31.

73 G. O. MacKenzie, *Transgender Nation* (Bowling Green, OH, 1994), p. 6.

74 R. Styles, *The New Girl: A Trans Girl Tells It Like It Is* (London, 2017), Ebook, locs. 111–16.

75 See the essays in E. Deshane (ed.), *#Trans: An Anthology About Transgender and Nonbinary Identity Online* (Santa Cruz, 2017).

76 T. Milan, 'For Years, Tiq Milan Felt Like the Only Black Trans Man on Earth', *Vice*, 27 March 2019: www.vice.com/en_us/article/zma9pe/tiq-milan-black-trans-man-community-online.

77 S. McGriever, 'The Mirror of Truth', in Deshane (ed.), *#Trans*, pp. 24–30, quote at p. 25.

78 H. Figa, 'YouTube Auto-Ethnography: An Introduction of Sorts', in Deshane (ed.), *#Trans*, pp. 14–22, quotes at pp. 14, 20.

79 M. Robertson, *Growing Up Queer: Kids and the Remaking of LGBTQ Identity* (New York, 2019), ch. 4.

80 Beemyn and Rankin, *Lives of Transgender People*, pp. 44–5, 54, 57–9, 75, 121, 140–1. Given that the survey was Internet based, these are hardly untainted findings.

81 Bornstein and Drucker, 'Gender is a Playground', 28.

82 J. Tobia, *Sissy: A Coming-Of-Gender Story* (New York, 2019), pp. 96, 165.

83 B. Bissinger, 'Across the Ages', *Vanity Fair: Trans America*, Special Edition, 18 August 2015.

84 J. Halberstam, *Trans*: A Quick and Quirky Account of Gender Variability* (Oakland, Calif., 2018).

85 Lester, *Trans Like Me*, loc. 2114.

86 R. Gossett, E. A. Stanley, and J. Burton (eds.), *Trap Door: Trans Cultural Production and the Politics of Visibility* (Cambridge, Mass., 2017).

87 J. Gill-Peterson, *Histories of the Transgender Child* (Minneapolis, Minn., 2018), ch. 2, quote at p. 11.

88 For heterosexuality before heterosexuality, and homosexuality before homosexuality, see K. M. Philips and B. Reay, *Sex Before Sexuality: A Premodern History* (Cambridge, 2011).

89 J. M. Irvine, *Disorders of Desire: Sexuality and Gender in Modern American Sociology* (Philadelphia, 2005), p. 208.

90 For an analysis of the meanings of trans, see R. Brubaker, *Trans: Gender and Race in an Age of Identities* (Princeton, NJ, 2016).

1 Before Trans

1 L. L. Stanley, *Men at Their Worst* (New York, 1940), p. 203.

2 E. Blue, 'The Strange Career of Leo Stanley: Remaking Manhood and Medicine at San Quentin State Penitentiary, 1913–1951', *Pacific Historical Review*, 78:2 (2009), 210–41.

3 L. L. Stanley, 'Experiences in Testicle Transplantation', *California State Journal of Medicine*, 43:7 (1920), 251–3; L. L. Stanley, 'Testicular Substance Implantation: Comments on Some Six Thousand Implantations', *California and Western Medicine*, 35:6 (1931), 411–15.

4 Stanley, *Men at Their Worst*, p. 203.

5 California Historical Society, Finding Aid to the Leo L. Stanley Scrapbooks and Papers, 1849–1974 (bulk 1928–1965), MS 2061: http://oac.cdlib.org/findaid/ark:/13030/c80863rn.

6 Stanley, *Men at Their Worst*, p. 203.

7 Ibid.

8 Ibid.

9 For a sensible introduction to this problem, see G. Beemyn, 'A Presence in the Past: A Transgender Historiography', *Journal of*

Women's History, 25:4 (2013), 113–21. See, too, Joanne Meyerowitz's introduction to the historical situation before transsexuality: J. Meyerowitz, *How Sex Changed: A History of Transsexuality in the United States* (Cambridge, MA, 2002), ch. 1; and F. Enke, 'Transgender History (and Otherwise Approaches to Queer Embodiment)', in D. Romesburg (ed.), *The Routledge History of Queer America* (New York, 2018), ch. 17.

10 A term used by Clare Sears, 'All that Glitters: Trans-ing California's Gold Rush Migrations', *GLQ*, 14:2–3 (2008), 383–402; C. Sears, *Arresting Dress: Cross-Dressing, Law, and Fascination in Nineteenth-Century San Francisco* (Durham, NC, 2014). See also P. Boag, *Re-Dressing America's Frontier Past* (Berkeley and Los Angeles, 2011).

11 E. Heaney, *The New Woman: Literary Modernism, Queer Theory, and the Transfeminine Allegory* (Evanston, Ill., 2017).

12 Ibid., pp. 169, 170, 172, 175.

13 J. Prosser, 'Transsexuals and the Transsexologists: Inversion and the Emergence of Transsexual Subjectivity', in L. Bland and L. Doan (eds.), *Sexology in Culture* (Chicago, 1998), ch. 7.

14 K. Amin, 'Glands, Eugenics, and Rejuvenation in *Man into Woman*: A Biopolitical Genealogy of Transsexuality', *TSQ*, 5:4 (2018), 589–605, quote at 593. This is a special issue of TSQ: *Trans*historicities*, ed. L. DeVun and Z. Tortorici.

15 R. H. Cleves, 'Six Ways of Looking at a Trans Man? The Life of Frank Shimer (1826–1901)', *Journal of the History of Sexuality*, 27:1 (2018), 32–62, quote at 35.

16 Ibid.

17 E. Skidmore, *True Sex: The Lives of Trans Men at the Turn of the 20th Century* (New York, 2017), quote at p. 2.

18 Sears, *Arresting Dress*, ch. 5.

19 Boag, *Re-Dressing America's Frontier Past*, p. 30.

20 For the variety of names for two-spirit people, see J. Roscoe, 'Glossary of Native Terms', in his *Changing Ones: Third and Fourth Genders in Native America* (New York, 1998), pp. 213–22.

21 C. Callender and others, 'The North American Berdache [and Comments and Reply]', *Current Anthropology*, 24:4 (1983), 443–70; Roscoe, 'Tribal Index', in *Changing Ones*, pp. 223–47.

71 G. Mak, 'Sandor/Sarolta Vay: From Passing Woman to Sexual Invert', *Journal of Women's History*, 16:1 (2004), 54–77, quote at 57. Mak uses the physician's account that Krafft-Ebing drew on.

72 Krafft-Ebing, *Psychopathia Sexualis*, p. 286.

73 Mak, 'Sandor/Sarolta Vay', 56.

74 Ibid., 56, 58.

75 Prosser, 'Transsexuals and the Transsexologists', pp. 124–5.

76 Krafft-Ebing, *Psychopathia Sexualis*, pp. 286, 289.

77 Ibid., pp. 285, 286, 288.

78 Ibid., p. 289.

79 Mak, 'Sandor/Sarolta Vay', 64.

80 Hirschfeld, *Transvestites*, p. 233.

81 Ibid., p. 124.

82 Ibid., p. 129.

83 Ibid., p. 34.

84 Ibid., p. 234.

85 Ibid., p. 129.

86 Ibid., p. 91.

87 Ibid., pp. 83, 93.

88 Ibid., p. 102.

89 Ibid., pp. 98, 102.

90 Ibid., p. 100.

91 D. B. Hill, 'Sexuality and Gender in Hirschfeld's *Die Transvestiten*: A Case of the "Elusive Evidence of the Ordinary"', *Journal of the History of Sexuality*, 14:3 (2005), 316–32.

92 Hirschfeld, *Transvestites*, pp. 56, 141 (for mention of police officer status).

93 Ibid., p. 63.

94 E. Gutheil, 'The Psychological Background of Transsexualism and Transvestism', *American Journal of Psychotherapy*, 8:2 (1954), 231–9, quote at 233.

95 Prosser, 'Transsexuals and the Transsexologists', p. 130.

96 Ibid., pp. 120–1.

97 Hirschfeld, *Transvestites*, p. 28.

98 Ibid., pp. 32, 34.

99 Ibid., p. 52.

100 Ibid., p. 84.

101 Ibid., p. 116.

102 Hill, 'Sexuality and Gender in Hirschfeld's *Die Transvestiten*', 324.

103 Hirschfeld, *Transvestites*, p. 31.

104 Ibid., p. 63.

105 Ibid., p. 78.

106 Ibid., pp. 34–5.

107 Ibid., p. 35.

108 Ibid.

109 H. Ellis, 'Sexo-Aesthetic Inversion', *The Alienist and Neurologist*, 24 (1913), 156–67, and 249–79; H. Ellis, 'Eonism', in his *Studies in the Psychology of Sex*, 2 vols. (New York, 1937), Vol. II, pp. 1–110, first published in 1928. See also the valuable essay by I. Crozier, 'Havelock Ellis, Eonism and the Patient's Discourse; or, Writing a Book About Sex', *History of Psychiatry*, 11 (2000), 125–54.

110 Ellis, 'Eonism', p. 1.

111 Ibid., p. 28.

112 Ibid.

113 Crozier, 'Havelock Ellis, Eonism and the Patient's Discourse', 131–2, 141.

114 Ibid., 133; Ellis, 'Eonism', p. 8.

115 H. Benjamin, *The Transsexual Phenomenon* (New York, 1966), p. 19.

116 Ellis, 'Sexo-Aesthetic Inversion', 164–7.

117 Ibid., 256–7.

118 Ibid., 256.

119 Ibid., 259.

120 Ibid., 261.

121 Ellis, 'Eonism', p. 62.

122 Ibid.

123 Ellis, 'Sexo-Aesthetic Inversion', 261, 269.

124 Ibid., 271.

125 Ibid., 272.

126 Ibid., 273.

127 Ellis, 'Eonism', p. 66.

128 Ibid., pp. 71, 74.

129 Ibid., p. 86.

130 Ibid., p. 83.

131 Ellis, 'Sexo-Aesthetic Inversion', 159.

132 D. J. Prickett, 'Magnus Hirschfeld and the Photographic (Re) Invention of the "Third Sex"', in G. Finney (ed.), *Visual Culture in Twentieth-Century Germany: Text as Spectacle* (Bloomington, Ind., 2006), ch. 7; K. Peters, 'Anatomy is Sublime: The Photographic Activity of Wilhelm Von Gloeden and Magnus Hirschfeld', in M. T. Thomas, A. F. Timm, and R. Herrn (eds.), *Not Straight from Germany: Sexual Publics and Sexual Citizenship Since Magnus Hirschfeld* (Ann Arbor, Mich., 2017), pp. 170–90; K. Sutton, 'Sexology's Photographic Turn: Visualizing Trans Identity in Interwar Germany', *Journal of the History of Sexuality*, 27:3 (2018), 442–79.

133 Quoted in R. Beachy, *Gay Berlin: Birthplace of Modern Identity* (New York, 2015), p. 178.

134 L. L. Lenz, *The Memoirs of a Sexologist: Discretion and Indiscretion* (New York, 1954), p. 463.

135 Ibid., p. 489.

136 Beachy, *Gay Berlin*, p. 179; Benjamin, *Transsexual Phenomenon*, p. 12.

137 H. Bauer, *The Hirschfeld Archives: Violence, Death, and Modern Queer Culture* (Philadelphia, 2017), p. 104.

138 Beachy, *Gay Berlin*, p. 179. For the Institute, see also Bauer, *Hirschfeld Archives*, ch. 4; and R. Herrn, M. T. Taylor, and A. F. Timm, 'Magnus Hirschfeld's Institute for Sexual Science: A Visual Sourcebook', in Thomas, Timm, and Herrn (eds.), *Not Straight from Germany*, pp. 37–79.

139 Prosser, 'Transsexuals and the Transsexologists', p. 121.

140 K. Sutton, 'Sexological Cases and the Prehistory of Transgender Identity Politics in Interwar Germany', in J. Damousi, B. Lang, and K. Sutton (eds.), *Case Studies and the Dissemination of Knowledge* (New York, 2015), ch. 5, quote at p. 85.

141 M. Hirschfeld, 'Transvestism', in *Sexual Anomalies: The Origins, Nature, and Treatment of Sexual Disorders* (New York, 1956), ch. 10, quote at p. 167. First published in 1948.

142 P. M. Wise, 'Case of Sexual Perversion', *The Alienist and Neurologist*, 4:1 (1883), 87–91.

143 Ibid., 89.

144 Ibid., 90.

145 Ibid.

146 Ibid., 89.

147 L. Faderman, 'Lucy Ann Lobdell', in M. Stein (ed.), *Encyclopedia of Lesbian, Gay, Bisexual and Transgendered History in America* (New York, 2004); J. M. Sloop, 'Lucy Lobdell's Queer Circumstances', in C. E. Morris (ed.), *Queering Public Address: Sexualities in American Historical Discourse* (Columbia, SC, 2007), pp. 149–73; B. L. Lobdell, *'A Strange Sort of Being': The Transgender Life of Lucy Ann / Joseph Israel Lobdell, 1829–1912* (Jefferson, NC, 2012). Lobdell is one of Skidmore's trans men: Skidmore, *True Sex*, pp. 27–36.

148 J. C. Shaw and G. N. Ferris, 'Perverted Sexual Instinct', *The Journal of Nervous and Mental Disease*, 10:2 (1883), 185–204, quotes at 188, 191, 197, 198.

149 Ibid., 203.

150 J. G. Kiernan, 'Sexual Perversion, and the Whitechapel Murders', *The Medical Standard*, 4:5 (1888), 129–30, quote at 130. For his citation of Lobdell (by description rather than by name), see J. G. Kiernan, 'Sexual Perversion, and the Whitechapel Murders (Continued)', *The Medical Standard*, 4:6 (1888), 170–2.

151 W. L. Howard, 'Psychical Hermaphroditism: A Few Notes on Sexual Perversion, with Two Clinical Cases of Sexual Inversion', *The Alienist and Neurologist*, 18:2 (1897), 111–18.

152 Ibid., 112.

153 Ibid., 115.

154 Ibid., 117.

155 C. W. Allen, 'Report of a Case of Psycho-Sexual Hermaphroditism', *Medical Record*, 51:9 (1897), 653–5.

156 Ibid., 655.

157 A. Flint, 'A Case of Sexual Inversion, Probably with Complete Sexual Anesthesia', *New York Medical Journal*, 2 December 1911.

158 A. Forel, *The Sexual Question: A Scientific, Psychological, Hygienic and Sociological Study*, ed. C. F. Marshall (New York, 1908), pp. 241, 248.

159 C. Hartland, *The Story of a Life: For the Consideration of the Medical Fraternity* (St Louis, Mo., 1901). The quotes come from the 1985 edition, with a foreword by C. A. Tripp: C. Hartland, *The Story of a Life* (San Francisco, 1985), Preface, and pp. 22, 59, 68, 72.

160 Ibid., Preface.

161 And interpreted as such: A. Sonstegard, 'Performing the "Unnatural" Life: America's First Gay Autobiography', *Biography*, 24:4 (2002), 545–68.

162 J. G. Kiernan, 'Sexology: Invert Marriages', *The Urologic and Cutaneous Review*, 18 (1914), 550. For Skidmore, see E. Skidmore, 'Ralph Kerwineo's Queer Body: Narrating the Scales of Social Membership in the Early Twentieth Century', *GLQ*, 20:1–2 (2014), 141–66; Skidmore, *True Sex*, index: Kerwineo, Ralph.

163 J. A. Gilbert, 'Homo-Sexuality and Its Treatment', *The Journal of Nervous and Mental Disease*, 52:4 (1920), 297–322, quote at 298. Boag discusses Hart in *Re-Dressing America's Frontier Past*, pp. 9–16.

164 Gilbert, 'Homo-Sexuality and Its Treatment', 313–14.

165 Ibid., 317.

166 Ibid., 321.

167 Ibid., 319.

168 Ibid., 320.

169 Ibid., 321, 322.

170 B. S. Talmey, 'Transvestism: A Contribution to the Study of the Psychology of Sex', *New York Medical Review*, 21 February 1914.

171 Ibid.

172 Ibid.

173 C. B. Horton and E. K. Clarke, 'Transvestism or Eonism: Discussion, With a Report of Two Cases', *The American Journal of Psychiatry*, 87:6 (1931), 1025–30.

174 L. L. London, 'Psychosexual Pathology of Transvestism', *The Urologic and Cutaneous Review*, 37 (1933), 600–4, quotes at 602, 604.

175 Ibid., 602.

176 Ibid., 603.

177 D. M. Olkon and I. C. Sherman, 'Eonism With Added Outstanding Psychopathic Features: A Unique Psychopathological Case', *The Journal of Nervous and Mental Disease*, 99:2 (1944), 159–67, quote at 164. Interestingly, however, the paper is called 'Eonism'.

178 Ibid., 163.

179 Ibid., 166.

180 D. O. Cauldwell, 'Psychopathia Transexualis', *Sexology: Sex Science Magazine* (December 1949), 274–80.

181 Benjamin, *Transsexual Phenomenon*, pp. 11–12.

182 Gutheil, 'Psychological Background'.

183 Hirschfeld, *Sexual Anomalies*.

184 R. W. Shufeldt, 'Biography of a Passive Pederast', *The American Journal of Urology and Sexology*, 13 (1917), 451–60.

185 Ibid., 458.

186 Ibid., 459–60.

187 Ibid., 451.

188 Ibid., 456, 457.

189 Ibid., 457.

190 Ibid., 452, 453.

191 Ibid., 454.

192 Ibid., 457.

193 Ibid.

194 R. Werther, *Autobiography of an Androgyne*, ed. S. Herring (New Brunswick, NJ, 2008), p. 120. First published in 1918.

195 Ibid., p. 80.

196 Ibid., p. 81.

197 A. Herrmann, 'The Androgyne as "Fairie": A Self-Authored Case History', in her *Queering the Moderns: Poses/Portraits/Performances* (New York, 2000), pp. 142–63, 176–7, quote at p. 148. See, also, J. Meyerowitz, 'Thinking Sex with an Androgyne', *GLQ*, 17:1 (2011), 97–105.

198 Werther, *Autobiography of an Androgyne*, pp. 8, 11.

199 Ibid., p. 11. That would have made their desires heterosexual, not homosexual.

200 Ibid., pp. 93, 97.

201 Ibid., p. 20.

202 Ibid., p. 25.

203 Ibid., p. 31.

204 Ibid., p. 45.

205 Ibid., p. 14.

206 S. R. Ullman, *Sex Seen: The Emergence of Modern Sexuality in America* (Berkeley and Los Angeles, 1997), ch. 3; N. A. Boyd, *Wide*

Open Town: A History of Queer San Francisco to 1965 (Berkeley and Los Angeles, 2003), pp. 30–2; D. Hurewitz, *Bohemian Los Angeles and the Making of Modern Politics* (Berkeley and Los Angeles, 2007), pp. 26–39.

207 Boyd, *Wide Open Town*, p. 36.

208 See Internet Archive: https://archive.org/details/RaeBourbon-78sCollection.

209 Iceberg Slim, *Pimp: The Story of My Life* (London, 2009), p. 58. First published in 1967. He was in Milwaukee, Wisconsin, at the time.

210 University of Chicago Library Special Collections Research Center, Ernest Watson Burgess Papers (hereafter Burgess), Box 98, Folder 11: 'Homosexual Materials': 'Leo. Age 18. Colored'.

211 D. K. Johnson, 'The Kids of Fairytown: Gay Male Culture on Chicago's Near North Side in the 1930s', in B. Beemyn (ed.), *Creating a Place for Ourselves: Lesbian, Gay, and Bisexual Histories* (New York, 1997), ch. 4, quote at p. 101.

212 G. Chauncey, *Gay New York: Gender, Urban Culture, and the Making of the Gay Male World, 1890–1940* (New York, 1994), ch. 11; Hurewitz, *Bohemian Los Angeles*, pp. 118–22; C. Heap, *Slumming: Sexual and Racial Encounters in American Nightlife, 1885–1940* (Chicago, 2009), ch. 6; J. F. Wilson, *Bulldaggers, Pansies, and Chocolate Babies: Performance, Race, and Sexuality in the Harlem Renaissance* (Ann Arbor, Mich., 2011), chs. 3, 5; J. Elledge, *The Boys of Fairy Town* (Chicago, 2018), ch. 12.

213 For Bentley, see A. Stavney, 'Cross-Dressing Harlem, Re-Dressing Race', *Women's Studies: An Interdisciplinary Journal*, 28:2 (1999), 127–56; Wilson, *Bulldaggers, Pansies, and Chocolate Babies*, ch. 5. For Winston, see M. Dancer, '"Gloria Swanson" Buried in Harlem', *The Chicago Defender*, 4 May 1940; Wilson, *Bulldaggers, Pansies, and Chocolate Babies*, pp. 194–5.

214 T. Russell, 'The Color of Discipline: Civil Rights and Black Sexuality', *American Quarterly*, 60:1 (2008), 101–28, quotes at 103.

215 C. McKay, *Home to Harlem* (Boston, 1987), p. 32. First published in 1928.

216 Ibid., p. 8.

217 A. B. C. Schwarz, *Gay Voices of the Harlem Renaissance* (Bloomington and Indianapolis, 2003), p. 104.

218 A. Y. Davis, *Blues Legacies and Black Feminism: Gertrude 'Ma' Rainey, Bessie Smith, and Billie Holiday* (New York, 1999), pp. 242–3.

219 K. Gallon, '"No Tears for Alden": Black Female Impersonators as "Outsiders Within" in the *Baltimore African American*', *Journal of the History of Sexuality*, 27:3 (2018), 367–94, quotes at 370.

220 R. Matthews, 'The Pansy Craze: Is It Entertainment Or Just Plain Filth?' *Baltimore Afro-American*, 6 October 1934.

221 P. Henderson, 'Photo News: Female Impersonators in Washington Show', *Baltimore Afro-American*, 25 September 1937. For Gallon, see 'Black Female Impersonators', 377.

222 Burgess, Box 98, Folder 11: 'Homosexual Materials': 'My Story of Fags, Freaks and Women Impersonators by Walt Lewis'.

223 Ibid.

224 Kinsey Institute, University of Indiana, Bloomington (hereafter KI), Thomas Painter, 'Male Homosexuals and Their Prostitutes in Contemporary America' (New York, 1941), Vol. II: 'The Prostitute', p. 183.

225 Werther, *Autobiography of an Androgyne*, pp. 120, 153.

226 Painter, 'Male Homosexuals', Vol. II, p. 184.

227 Ibid., p. 185.

228 Ibid., pp. 185–6.

229 Ibid., p. 185.

230 B. Reay, *New York Hustlers: Masculinity and Sex in Modern America* (Manchester, 2010), p. 59.

231 Boyd, *Wide Open Town*, pp. 25–38, quote at p. 35.

232 For Cadmus, see J. Weinberg, 'Cruising with Paul Cadmus', *Art in America*, 80:11 (1992), 102–9; R. Meyer and A. D. Weinberg, *Paul Cadmus: The Sailor Trilogy* (New York, 1996) (an exhibition catalogue for the Whitney Museum of American Art); and R. Meyer, 'A Different American Scene: Paul Cadmus and the Satire of Sexuality', in his *Outlaw Representation: Censorship and Homosexuality in Twentieth-Century American Art* (Boston, 2002), ch. 2.

233 G. W. Henry, *Sex Variants: A Study of Homosexual Patterns*, 2 vols. (New York, 1941), Vol. I, p. 440.

234 Ibid., p. 446.

235 Ibid., p. 447.

236 Ibid., pp. 414, 422.

237 Ibid., p. 432.

238 Ibid, pp. 432–3.

239 Ibid., pp. 435, 438.

240 Ibid., pp. 538, 542.

241 Ibid., p. 540.

242 Ibid., p. 541.

243 Ibid., p. 542.

244 Talmey, 'Transvestism'.

245 Henry, *Sex Variants*, Vol. I, p. 497.

246 Ibid., pp. 492, 497, 498.

247 Ibid., pp. 492, 494.

248 For the photographs, see Henry, *Sex Variants*, Vol. II, pp. 1053–5.

249 Henry, *Sex Variants*, Vol. I, p. 495. For Eugen Steinach, see C. Sengoopta, '"Dr Steinach Coming to Make Old Young!": Sex Glands, Vasectomy and the Quest for Rejuvenation in the Roaring Twenties', *Endeavour*, 27:3 (2003), 122–6. For Benjamin as a practitioner, see M. A. Kozminski and D. A. Bloom, 'A Brief History of Rejuvenation Operations', *The Journal of Urology*, 187:3 (2012), 1130–4, at 1132.

250 Henry, *Sex Variants*, Vol. I, p. 495.

251 For example, Burgess, Box 98, Folder 2: 'Homosexuality Interviews'; Box 98, Folder 11: 'Homosexual Materials'.

252 Burgess, Box 98, Folder 11: 'Homosexual Materials': 'Jokes'; 'Queer Jokes'.

253 Burgess, Box 145, Folder 8: 'Glossary of Homosexual Terms'.

254 See, for example, H. Benjamin, 'The Steinach Operation: Report of 22 Cases with Endocrine Interpretation', *Endocrinology*, 6:6 (1922), 776–86; P. Schmidt, *The Theory and Practice of the Steinach Operation with a Report on One Hundred Cases* (London, 1924). For a medical-historical overview, see P. Södersten and others, 'Eugen Steinach: The First Neuroendocrinologist', *History of Endocrinology*, 155:3 (2014), 688–702. For the cultural impact, see M. Pettit, 'Becoming Glandular: Endocrinology, Mass Culture, and Experimental Lives in the Interwar Age', *American Historical Review*, 118:4 (2013),

1052–76; M. Makela, 'Rejuvenation and Regen(d)eration: *Der Steinachfilm*, Sex Glands, and Weimar-Era Visual and Literary Culture', *German Studies Review*, 38:1 (2015), 35–62.

255 Quoted in D. Serlin, *Replaceable You: Engineering the Body in Postwar America* (Chicago, 2004), p. 127.

256 G. Bentley, 'I Am a Woman Again', *Ebony* (August 1952). Serlin discusses the case in *Replaceable You*, ch. 3: 'Gladys Bentley and the Cadillac of Hormones'.

257 J. Meyerowitz, 'Sex Change and the Popular Press: Historical Notes on Transsexuality in the United States, 1930–1955', *GLQ*, 4:2 (1998), 159–87, quote (from 1934) at 166.

258 Pagan Kennedy discusses both cases in *The First Man-Made Man: The Story of Two Sex Changes, One Love Affair, and a Twentieth-Century Medical Revolution* (New York, 2007). See, also, Dillon's autobiography: M. Dillon / L. Jivaka, *Out of the Ordinary: A Life of Gender and Spiritual Transitions*, ed. J. Lau and C. Partridge (New York, 2017).

259 Kennedy, *First Man-Made Man*, p. 5.

260 Ibid., p. 17.

261 Ibid., pp. 14–15, 17. See M. Dillon, *Self: A Study in Ethics and Endocrinology* (London, 1946).

262 Dillon, *Self*, pp. 52, 53.

263 Ibid., p. 53.

264 Ibid., p. 50.

265 Kennedy, *First Man-Made Man*, pp. 91–2.

266 Ibid., ch. 7.

267 H. Gillies and D. R. Millard, *The Principles and Art of Plastic Surgery*, 2 vols. (London, 1957), Vol. II, p. 387.

268 A. McIndoe, 'The Treatment of Congenital Absence and Obliterative Conditions of the Vagina', *British Journal of Plastic Surgery*, 2:4 (1950), 254–67.

269 E. de Savitsch, *Homosexuality, Transvestism and Change of Sex* (London, 1958), pp. 54–7.

270 Ibid, pp. 61–77, 96–118.

271 Ibid., p. 116.

272 Ibid., p. 97.

273 See N. Hoyer (ed.), *Man Into Woman: An Authentic Record of a*

Change of Sex, trans. H. J. Stenning (London, 1933). For Amin, see Amin, 'Glands, Eugenics, and Rejuvenation'.

274 G. Mak, *Doubting Sex: Inscriptions, Bodies and Selves in Nineteenth-Century Hermaphrodite Case Histories* (Manchester, 2012), esp. ch. 2.

275 Ibid., ch. 7; C. Matta, 'Ambiguous Bodies and Deviant Sexualities: Hermaphrodites, Homosexuality, and Surgery in the United States, 1850–1904', *Perspectives in Biology and Medicine*, 48:1 (2005), 74–83; E. Reis, *Bodies in Doubt: An American History of Intersex* (Baltimore, 2009), chs. 3–4. For some examples, see H. H. Young, 'Some Hermaphrodites I Have Met', *New England Journal of Medicine*, 209:8 (1933), 370–5; H. H. Young, *Genital Abnormalities, Hermaphroditism and Related Adrenal Diseases* (Baltimore, 1937).

276 J. R. Goffe, 'A Pseudohermaphrodite in which the Female Characteristics Predominated; Operation for Removal of the Penis and the Utilization of the Skin Covering it for the Formation of a Vaginal Canal', *American Journal of Obstetrics and Diseases of Women and Children*, 48:6 (1903), 755–63.

277 Matta, 'Ambiguous Bodies'; Reis, *Bodies in Doubt*, ch. 3.

278 W. Martin, 'Pseudohermaphrodotismus Masculinis', *Surgical Clinics of North America*, 9 (1929), 535–44, quote at 542.

279 Young, *Genital Abnormalities*, pp. 139–42, quotes at p. 142. The case is discussed sensitively in J. Gill-Peterson, 'Trans of Color Critique before Transsexuality', *TSQ*, 5:4 (2018), 606–20.

280 Meyerowitz, *How Sex Changed*, p. 15. Emphasis in original.

281 D. A. Griffiths, 'Diagnosing Sex: Intersex Surgery and "Sex Change" in Britain 1930–1955', *Sexualities*, 21:3 (2018), 476–95. Griffiths discusses Weston, Dillon, and Cowell.

282 Ibid., 477.

283 J. Gill-Peterson, *Histories of the Transgender Child* (Minneapolis, Minn., 2018), ch. 2, quote at p. 89.

284 Ibid., p. 95.

285 Meyerowitz, 'Thinking Sex with an Androgyne', 102.

2 The Transsexual Moment

1 See H. Benjamin, 'Introduction', in R. Green and J. Money (eds.), *Transsexualism and Sex Reassignment* (Baltimore, 1969), pp.

1–10, though he acknowledges Cauldwell's role: 'Whether I had ever read that article and the expression had remained in my subconscious, frankly, I do not know' (at p. 4).

2 H. Benjamin, 'Transsexualism and Transvestism as Psycho-Somatic and Somatic-Psychic Syndromes', *American Journal of Psychotherapy*, 8:2 (1954), 219–30, quote at 220.

3 Ibid.

4 Ibid.

5 J. Money, *Gay, Straight, and In-between: The Sexology of Erotic Orientation* (New York, 1988), p. 88.

6 D. O. Cauldwell, 'Psychopathia Transexualis', *Sexology: Sex Science Magazine*, 16:5 (1949), 274–80.

7 D. O. Cauldwell, *Questions and Answers on the Sex Life and Sexual Problems of Trans-Sexuals* (Girard, Kans., 1950).

8 C. Hamburger, G. K. Stürup, and H. E. Dahl-Iversen, 'Transvestism: Hormonal, Psychiatric, and Surgical Treatment', *Journal of the American Medical Association*, 152:5 (1953), 391–6.

9 D. O. Cauldwell (ed.), *Transvestism . . . Men in Female Dress* (New York, 1956).

10 N. Haire, 'Is Change of Sex Possible?' *Sexology: Sex Science Magazine*, 18:7 (1952), 420–7.

11 Observations from *Sexology*'s Letters to the Editor section during the 1950s. Cauldwell edited this section.

12 J. Meyerowitz, *How Sex Changed: A History of Transsexuality in the United States* (Cambridge, MA, 2002), chs. 1 and 2.

13 Ibid., ch. 4. The quote comes from the title of the chapter.

14 C. Hamburger, 'The Desire for Change of Sex as Shown by Personal Letters from 465 Men and Women', *Acta Endocrinologica*, 14 (1953), 361–75.

15 K. M. Bowman and B. Engle, 'Medicolegal Aspects of Transvestism', *The American Journal of Psychiatry*, 113:7 (1957), 583–8.

16 J. J. Hage, R. B. Karim, and D. R. Laub, 'On the Origin of Pedicled Skin Inversion Vaginoplasty: Life and Work of Dr Georges Burou of Casablanca', *Annals of Plastic Surgery*, 59:6 (2007), 723–9.

17 'Dear Abbé: Advice to the TV-Lorn', *Turnabout: A Magazine of Transvestism*, 1 (1963), 20; S. Fredericks, 'Kaleidoscope', *Turnabout: A Magazine of Transvestism*, 2 (1963), 30.

18 Letters to the Editor: 'Change of Sex', *Sexology: Sex Science Magazine*, 32:11 (1966), 764–5.

19 Bowman and Engle, 'Medicolegal Aspects of Transvestism', 588.

20 Ibid., 587.

21 Cauldwell, *Questions and Answers*, p. 7.

22 Letters to the Editor: 'Wants to be a Woman', *Sexology: Sex Science Magazine*, 16:7 (1950), 457–8.

23 Letters to the Editor: 'Change of Sex', *Sexology: Sex Science Magazine*, 21:6 (1955), 389–90; Letters to the Editor: 'Male Transvestite', *Sexology: Sex Science Magazine*, 21:10 (1955), 666–7.

24 F. G. Worden and J. T. Marsh, 'Psychological Factors in Men Seeking Sex Transformation: A Preliminary Report', *Journal of the American Medical Association*, 157:15 (1955), 1292–8, quote at 1292.

25 Ibid., 1293–4.

26 Ibid., 1295.

27 Ibid., 1298.

28 J. Meyerowitz, 'Sex Research at the Borders of Gender: Transvestites, Transsexuals, and Alfred C. Kinsey', *Bulletin of the History of Medicine*, 75:1 (2001), 72–90, esp. 74–80. Lawrence is also discussed in Robert Hill's thesis: R. S. Hill, '"As a Man I Exist; As a Woman I Live": Heterosexual Transvestism and the Contours of Gender and Sexuality in Postwar America', University of Michigan Ph.D., 2007, pp. 49–52, 54.

29 Meyerowitz, 'Sex Research at the Borders of Gender', 89.

30 Kinsey Institute, University of Indiana, Bloomington (hereafter KI), The Harry Benjamin Collection (hereafter Benjamin), Box 3, Series 2c, Correspondence, Folder: B., V. B., Letters: 19 July 1948, 16 Nov. 1949, 22 Nov. 1949.

31 Ibid., Letter: 19 July 1948.

32 Ibid., Letter. 31 May 1949.

33 Ibid., Letters: 30 Sept. 1949, 5 Dec. 1949.

34 Ibid., Letter: 2 Oct. 1950.

35 Ibid., Letter: 22 Sept. 1953.

36 Benjamin, Box 6, Series 2c, Correspondence, Folder: J., R. (C.) 1954–7.

37 Benjamin, Box 28, Series 6g, Additional Materials, Folder 20: Master List of All Patients.

38 Benjamin, Box 3, Series 2c, Correspondence, Folder: Belt, Dr Elmer, 1962–5, Letter: 29 June 1962.

39 Ibid., 1959–62, Letter: 22 May 1961.

40 Ibid., 1962–5, Letter: 21 Dec. 1964.

41 Ibid., 1959–62, Letters: 13 Feb. 1961, 22 Dec. 1961; 1962–5, Letters: 14 Dec. 1964, 18 Dec. 1964, 29 Dec. 1964.

42 Ibid., 1959–62, Letter: 13 Feb. 1961.

43 Ibid., 1958–9, Letter: 30 Dec. 1958.

44 Ibid., Letter: 9 March 1959.

45 P. Morgan, *The Man-Maid Doll* (Secaucus, NJ, 1973), pp. 43, 46.

46 Benjamin, Box 3, Series 2c, Correspondence, Folder: Belt, Dr Elmer, 1959–62, Letter: 9 Jan. 1961.

47 Ibid.

48 Ibid., Letter: 22 Feb. 1960.

49 Ibid.

50 Ibid., Letter: 7 March 1960.

51 Ibid., 1962–5, Letters: 23 July 1962, 30 July 1962.

52 Ibid., Letter: 27 August 1962.

53 Ibid., Letter: 18 July 1964.

54 Ibid., 1965–71, Letters: 25 Jan. 1965, 8 Feb. 1965, 1 Feb. 1965.

55 Ibid., Letter: 3 Sept. 1968.

56 Benjamin, Box 25, Series 6c, Additional Materials, Folder 11: Letter: 15 April 1963.

57 Benjamin, Box 3, Series 2c, Correspondence, Folder: Belt, Dr Elmer, 1958–9, Letter: 9 Dec. 1958; 1965–71, Letter: 8 Dec. 1967.

58 Ibid., 1958–9, Letters: 1 Dec. 1958, 3 Dec. 1958, 9 March 1959.

59 Ibid., 1962–5, Letter: 16 Oct. 1962.

60 Ibid., 1965–71, Letter: 25 Oct. 1966.

61 Ibid., 1958–9, Letter: 28 July 1958.

62 Ibid., Letter: 11 June 1958.

63 I. B. Pauly, 'Male Psychosexual Inversion: Transsexualism: A Review of 100 Cases', *Archives of General Psychiatry*, 13 (1965), 172–81, quote at 172.

64 Ibid., 172.

65 Ibid., 179.

66 I. B. Pauly, 'The Current Status of the Change of Sex Operation', *The Journal of Nervous and Mental Disease*, 147:5 (1968), 460–71, quote at 460.

67 Letters to the Editor: 'Change of Sex', *Sexology: Sex Science Magazine*, 30:9 (1964), 614.

68 Ibid.

69 Letters to the Editor: 'Sex-Change Operation', *Sexology: Sex Science Magazine*, 31:3 (1964), 186.

70 Letters to the Editor: 'Sex Change', *Sexology: Sex Science Magazine*, 32:5 (1965), 329.

71 Letters to the Editor: 'Woman Into Man', *Sexology: Sex Science Magazine*, 35:8 (1969), 546.

72 A. Sinclair, *I Was Male!* (Chicago, 1965), cover quote.

73 Ibid., pp. 9, 17.

74 H. J. Star, *My Unique Change* (Chicago, 1965), p. 9.

75 Ibid.

76 M. A. Costa, *Reverse Sex* (London, 1965), p. 9.

77 Ibid., p. 8.

78 F. L. Shaw, 'A Plea for Skepticism', *Turnabout: A Magazine of Transvestism*, 2 (1963), 2.

79 S. Fredericks, 'Kaleidoscope', *Turnabout: A Magazine of Transvestism*, 2 (1963), 28–9.

80 H. Benjamin, 'Advice to a Transsexual', *Turnabout: A Magazine of Transvestism*, 3 (1964), 7–10.

81 'Dear Abbé': Letter from Harry Benjamin, *Turnabout: A Magazine of Transvestism*, 2 (1963), 19; H. Benjamin, *The Transsexual Phenomenon* (New York, 1966), p. 38.

82 H. Benjamin, 'Abstract: Clinical Aspects of Transsexualism in the Male and Female', *Turnabout: A Magazine of Transvestism*, 2 (1963), 10–11.

83 S. Fredericks, 'Kaleidoscope', *Turnabout: A Magazine of Transvestism*, 3 (1964), 21–3.

84 L. Channing, 'Each Day I Live a Lie', *Turnabout: A Magazine of Transvestism*, 3 (1964), 11; Benjamin, *Transsexual Phenomenon*, pp. 69–70.

85 Benjamin, Box 14, Series 3a, Clinical/Research: Patient Files, Folder 15: Meetings of Female [*sic*] TS Groups, 9 March 1969.

86 Ibid., Folder 16: Meetings of Male [*sic*] TS Groups, 24 Nov. 1968, 12 Jan. 1969, 16 Feb. 1969.

87 Ibid., 12 Jan. 1969.

88 Ibid., 29 Sept. 1968.

89 Benjamin, Box 4, Series 2c, Correspondence, Folder: D., A. (A. S.), Letters: 13 June 1955, and undated letter.

90 D. King, 'Gender Blending: Medical Perspectives and Technology', in R. Ekins and D. King (eds.), *Blending Genders: Social Aspects of Cross-Dressing and Sex-Changing* (London, 1996), ch. 7, quote at p. 93.

91 J. K. Meyer and D. J. Reter, 'Sex Reassignment', *Archives of General Psychiatry*, 36:9 (1979), 1010–15, quote at 1010.

92 The best recent history is K. Karkazis, *Fixing Sex: Intersex, Medical Authority, and Lived Experience* (Durham, NC, 2008).

93 R. J. Stoller, *Sex and Gender: On the Development of Masculinity and Femininity* (New York, 1968), chs. 5–7.

94 For Johns Hopkins and Money and intersex, see Karkazis, *Fixing Sex*, ch. 2.

95 A. H. Devor and N. Matte, 'ONE Inc. and Reed Erickson: The Uneasy Collaboration of Gay and Trans Activism, 1964–2003', *GLQ*, 10:2 (2004), 179–209; A. Devor and N. Matte, 'Building a Better World for Transpeople: Reed Erickson and the Erickson Educational Foundation', *International Journal of Transgenderism*, 10:1 (2007), 47–68.

96 S. R. Wolf and others, 'Psychiatric Aspects of Transsexual Surgery Management', *The Journal of Nervous and Mental Disease*, 147:5 (1968), 525–31; J. K. Meyer, 'Clinical Variants Among Applicants for Sex Reassignment', *Archives of Sexual Behavior*, 3:6 (1974), 527–58; D. R. Laub and N. Fisk, 'A Rehabilitation Program for Gender Dysphoria Syndrome by Surgical Sex Change', *Plastic and Reconstructive Surgery*, 53:4 (1974), 388–403; E. L. Weitzman, C. A. Shamoian, and N. Golosow, 'Identity Diffusion and the Transsexual Resolution', *The Journal of Nervous and Mental Disease*, 151:5 (1970), 295–302; L. E. Newman, 'Transsexualism in Adolescence: Problems in Evaluation and Treatment', *Archives of General Psychiatry*, 23 (1970), 112–21; M. J. Pearson, 'A Diagnostic Survey of 23 Patients Applying for Gender Surgery', in D. R.

Laub and P. Gandy (eds.), *Proceedings of the Second Interdisciplinary Symposium on Gender Dysphoria Syndrome* (Stanford, 1973), pp. 105–14; A. J. Arieff, 'Five-Year Studies of Transsexuals: Psychiatric, Psychological and Surgical Aspects', in Laub and Gandy (eds.), *Proceedings of the Second Interdisciplinary Symposium*, pp. 240–3.

97　D. W. Hastings, 'Postsurgical Adjustment of Male Transsexual Patients', *Clinics in Plastic Surgery*, 1:2 (1974), 335–44; T. Kando, 'Passing and Stigma Management: The Case of the Transsexual', *The Sociological Quarterly*, 13:4 (1972), 475–83, esp. 476, n. 4.

98　Benjamin, Box 25, Folder 4: Transsexuals California, 1970.

99　Meyer, 'Clinical Variants Among Applicants for Sex Reassignment', 528.

100　J. K. Meyer, 'The Theory of Gender Identity Disorders', *Journal of the American Psychoanalytic Association*, 30:2 (1982), 381–418, quote at 381.

101　L. M. Lothstein, 'Countertransference Reactions to Gender Dysphoric Patients: Implications for Psychotherapy', *Psychotherapy: Theory, Research and Practice*, 14:1 (1977), 21–31; L. M. Lothstein, *Female-to-Male Transsexualism: Historical, Clinical and Theoretical Issues* (Boston, Mass., 1983), p. 17.

102　I. M. Dushoff, 'Economic, Psychologic and Social Rehabilitation of Male and Female Transsexuals Prior to Surgery', in Laub and Gandy (eds.), *Proceedings of the Second Interdisciplinary Symposium*, pp. 197–203.

103　Meyerowitz, *How Sex Changed*, pp. 217–22.

104　Department of Special Collections, Charles E. Young Research Library, UCLA, Robert J. Stoller Papers (hereafter Stoller), Box 37: Folder General Correspondence A–G, 1969–70, R. Stoller to H. B., 12 March 1970.

105　Benjamin, Box 28, Series 6g, Additional Materials, Folder 20: Master List of All Patients.

106　S. Churcher, 'The Anguish of the Transsexuals', *New York*, 18 June 1980.

107　D. H. Feinbloom, *Transvestites & Transsexuals: Mixed Views* (New York, 1976), p. 184.

108　Stoller, Box 38, Folder: General Correspondence H–P, 1975–6, R. Stoller to J. P., 27 May 1976.

109 Benjamin, Box 25, Folder 4: Transsexuals California, 1970.

110 S. Fredericks, 'Kaleidoscope', *Turnabout: A Magazine of Transvestism*, 1 (1963), 17.

111 Letters to the Editor: 'Transsexual', *Sexology: Sex Science Magazine*, 31:6 (1965), 395–6.

112 Churcher, 'The Anguish of the Transsexuals'.

113 Benjamin, *Transsexual Phenomenon*, p. 147.

114 G. M. Warner and M. Lahn, 'A Case of Female Transsexualism', *Psychiatric Quarterly*, 44:1–4 (1970), 467–87.

115 J. E. Hoopes, N. J. Knorr, and S. R. Wolf, 'Transsexualism: Considerations Regarding Sexual Reassignment', *The Journal of Nervous and Mental Disease*, 147:5 (1968), 510–16.

116 Archives & Special Collections, Augustus C. Long Health Sciences Library, Columbia University Medical Center, New York, Ethel Spector Person Papers, Box 14, Folder 1: Transsexualism Case History: Female-to-Male, S. K., 1974.

117 I. B. Pauly, 'Female Transsexualism: Part I', *Archives of Sexual Behavior*, 3:6 (1974), 487–507; I. B. Pauly, 'Female Transsexualism: Part II', *Archives of Sexual Behavior*, 3:6 (1974), 509–26.

118 Pauly, 'Female Transsexualism: Part I', 493–6.

119 Ibid., 499.

120 Ibid., 505.

121 E. McCauley and A. A. Ehrhardt, 'Follow-up of Females with Gender Identity Disorders', *The Journal of Nervous and Mental Disease*, 172:6 (1984), 353–8.

122 Ibid., 356, 357.

123 L. R. Derogatis, J. K. Meyer, and Patricia Boland, 'A Psychological Profile of the Transsexual: II. The Female', *The Journal of Nervous and Mental Disease*, 169:3 (1981), 157–68, quote at 166.

124 Ibid., 166.

125 Ibid., 165.

126 Lothstein, *Female-to-Male Transsexualism*, p. 9.

127 San Francisco History Center, San Francisco Public Library, Louis Graydon Sullivan Papers (hereafter Sullivan Papers), Box 1, Folders 4–20: Sheila/Louis Sullivan Diaries and Journals (hereafter Sullivan Diaries), 1961–91. See also L. M. Rodemeyer, *Lou Sullivan Diaries (1970–1980) and Theories of Sexual Embodiment:*

Making Sense of Sensing (New York, 2017), which contains extracts from the 1970–80 diaries.

128 Sullivan Diaries, 23 March 1973.

129 Ibid., 7 July 1973.

130 Ibid., 13 July 1973. Emphasis in original.

131 Ibid., 22 March 1976. Emphasis in original.

132 Ibid., 5 and 21 August 1976.

133 Sullivan Papers, Box 1, Folder 1: Application to Stanford Gender Dysphoria Program, October 1976.

134 Feinbloom, *Transvestites & Transsexuals*, p. 31.

135 Sullivan Diaries, 26 November 1976.

136 Sullivan Papers, Box 2, Folder 110: Letter to E. M., 15 August 1977.

137 Sullivan Papers, Box 2, Folder 44: Letter to Mario Martino [Angelo Tornabene], 12 September 1979. Emphasis in original.

138 Ibid., Folder 49B, Letter to M. B., 2 October 1979.

139 D. G. Raynor, *A Year Among the Girls* (London, 1967), p. 90. First published in New York in 1966.

140 Ibid., pp. 88–9.

141 Ibid., p. 124.

142 Ibid., p. 145.

143 V. Prince, 'The Art of Female Impersonation', *Transvestia*, 11 (1961), pp. 57–70.

144 'Susanna Says', *Transvestia*, 23 (1963), 70–6, quotes at 72–3.

145 Hill, 'Heterosexual Transvestism', p. 6.

146 M. Hurst and R. Swope (eds.), *Casa Susanna* (New York, 2014). First published in 2005.

147 J. T. Talamini, *Boys Will Be Girls: The Hidden World of the Heterosexual Male Transvestite* (Lanham, Md., 1982), p. 42.

148 B. Davis, 'Using Archives to Identify the Trans* Women of Casa Susanna', *TSQ*, 2:4 (2015), 621–34, quote at 625.

149 Hill, 'Heterosexual Transvestism', p. 6.

150 S. Hackett, 'Casa Susanna', in A. Pardo (ed.), *Another Kind of Life: Photography on the Margins* (London, 2018), p. 80. For the photographs, see pp. 59–69.

151 Ibid., p. 80.

152 Pardo (ed.), *Another Kind of Life*.

153 Hill, 'Heterosexual Transvestism', p. 12.

154 Hurst and Swope (eds.), *Casa Susanna*, no pagination.

155 Hill, 'Heterosexual Transvestism', p. 25.

156 A comparison made by ibid., p. 269.

157 The interviews are used (sparingly) in R. F. Docter, *From Man to Woman: The Transgender Journey of Virginia Prince* (Northridge, Calif., 2004), ch. 7.

158 H. G. Beigel, 'A Week in Alice's Wonderland', *The Journal of Sex Research*, 5:2 (1969), 108–22, quotes at 120.

159 Hill, 'Heterosexual Transvestism', pp. 64–6.

160 Hurst and Swope (eds.), *Casa Susanna*, no pagination. It has to be said that he did not at first understand the context of the Casa Susanna photographs, and, ironically, saw them in terms of 'Queendom', as conservatively dressed drag queens: 'a drag queen on an ugly sofa . . . happily knitting while dressed in conservative women's daywear'.

161 Raynor, *Year Among the Girls*, pp. 30–1.

162 C. Channing, 'Views/Reviews', *Turnabout: A Magazine of Transvestism*, 2 (1963), 22. She misspells Rechy as Reichy throughout the review.

163 S. Fredericks, 'Views/Reviews', *Turnabout: A Magazine of Transvestism*, 2 (1963), 23.

164 V. L. Bullough and B. Bullough, 'The Emergence of Organized Transvestism and Its Implications', in their *Cross Dressing, Sex, and Gender* (Philadelphia, 1993), ch. 12, quote at pp. 289–90.

165 H. G. Beigel and R. Feldman, 'The Male Transvestite's Motivation in Fiction, Research, and Reality', *Advances in Sex Research*, 1 (1963), 198–200, esp. 199.

166 R. J. Stoller, 'Pornography and Perversion', *Archives of General Psychiatry*, 22:6 (1970), 490–9, at 499.

167 See 'Purpose of Transvestia', *Transvestia*, 61 (1970), frontispiece.

168 'Editorial', 'Transvestia: Journal of the American Society for Equality in Dress', 1:1 (1952), 4.

169 Raynor, *Year Among the Girls*, pp. 51, 81.

170 Stoller, 'Pornography and Perversion', 495.

171 F. L. Shaw, 'Views/Reviews', *Turnabout: A Magazine of Transvestism*, 2 (1964), 26–7.

172 L. Maddock, *Sex Life of a Transvestite* (Los Angeles, 1964), p. 14.

173 Ibid., pp. 128–9.

174 H. Magnus, *Transvestism* (New York, 1969), p. 35.

175 Ibid., pp. 36–8.

176 Nutrix Corp., *Letters from Female Impersonators, Volume Number Fifteen* (New York, 1964).

177 The copyright entries are now available on Google Books. The details of Klaw's prosecution and successful appeal are in *United States* v. *Irving Klaw and Jack Kramer*, 350 F.2d 155 (2d Cir. 1965), Court of Appeals for the Second Circuit: www.courtlistener.com/opinion/268980/united-states-v-irving-klaw-and-jack-kramer.

178 E. Schaefer, 'The Obscene Seen: Spectacle and Transgression in Postwar Burlesque Films', *Cinema Journal*, 36:2 (1997), 41–66, quote at 59.

179 Ibid., 42.

180 Minette, *Recollections of a Part-Time Lady* (New York, 1979), no pagination.

181 D. Paulson and R. Simpson, *An Evening in the Garden of Allah: A Gay Cabaret in Seattle* (New York, 1996), p. 17.

182 Ibid., pp. 38, 78, 122, 127.

183 Ibid., pp. 24, 36.

184 Ibid., p. 77.

185 L. Grantmyre, '"They Lived Their Life and They Didn't Bother Anybody": African American Female Impersonators and Pittsburgh's Hill District, 1920–1960', *American Quarterly*, 63:4 (2011), 983–1011. For the photographs, see Carnegie Museum of Art, Pittsburgh, Charles 'Teenie' Harris Archive: http://teenie.cmoa.org/Default.aspx. The photographer was a man called Walter Allen rather than Harris himself (Grantmyre, 'They Lived Their Life', 1008, n. 22).

186 T. Russell, 'The Color of Discipline: Civil Rights and Black Sexuality', *American Quarterly*, 60:1 (2008), 101–28.

187 'Female Impersonators Hold Costume Balls', *Ebony*, March 1952.

188 Ibid.

189 Ibid.

190 A. Drexel, 'Before Paris Burned: Race, Class, and Male Homosexuality on the Chicago South Side, 1935–1960', in

B. Beemyn (ed.), *Creating a Place for Ourselves: Lesbian, Gay, and Bisexual Histories* (New York, 1997), ch. 5, quote at p. 134.

191 *Jet*, 8 September 1966; *Jet*, 16 November 1967. For Sir Lady Java, see T. Ellison, 'The Labor of Werqing It: The Performance and Protest Strategies of Sir Lady Java', in R. Gossett, E. A. Stanley, and J. Burton (eds.), *Trap Door: Trans Cultural Production and the Politics of Visibility* (Cambridge, MA, 2017), pp. 1–21; and 'Sir Lady Java', *Transas City*: http://transascity.org/sir-lady-java/.

192 G. Beemyn, *A Queer Capital: A History of Gay Life in Washington, D.C.* (New York, 2015), p. 111.

193 Russell, 'Color of Discipline', 106–7.

194 Ralston Crawford Collection, William Ransom Hogan Jazz Archive, Howard-Tilton Memorial Library, Tulane University. See Tulane University Digital Library: https://digitallibrary.tulane.edu/islandora/search/dew%20drop%20inn?type=dismax&islandora_solr_search_navigation=0&f%5B0%5D=mods_subject_topic_s%3A%22Bars%22.

195 See S. Leigh, 'Obituary: Bobby Marchan', *The Independent*, 18 December 1999; J. Yiannis, 'Bobby Marchan', *GayCultureLand*, 26 April 2017: http://gaycultureland.blogspot.com/2017/04/bobby-marchan.html; J. D. Doyle, 'Patsy Vadalia, also Valdelar, Vidalia', *Queer Music Heritage*, no date: http://queermusicheritage.com/drag-vidalia.html.

196 M. Hamilton, 'Sexual Politics and African-American Music: Or, Placing Little Richard in History', *History Workshop Journal*, 46 (1998), 160–76, quote at 162.

197 See J. D. Doyle, 'Club My-O-My . . . New Orleans', *Queer Music Heritage*, no date: http://queermusicheritage.com/fem-myomy.html; J. D. Doyle, 'Jimmy Callaway', *Queer Music Heritage*, no date: http://queermusicheritage.com/drag-callaway.html.

198 E. Schweigershausen, '"Private Birthday Party": Rare Photos From Kansas City's 1960s Drag Scene', *The Cut* [*New York Magazine*], 8 April 2014: www.thecut.com/2014/04/rare-photos-from-the-60s-kansas-city-drag-scene.html; K. Deel, 'LGBT In KC, A History', *Squeezebox*, 15 August 2015: www.squeezeboxcity.com/lgbt-in-kc-a-history.

199 For Arnold and Newton, see E. Newton, *Mother Camp:*

Female Impersonators in America (Englewood Cliffs, NJ, 1972), Acknowledgments, and pp. 69–88.

200 See the University of Missouri–Kansas City's Robert Heishman Collection: https://dl.mospace.umsystem.edu//umkc/islandora/object/umkc:heishman; and the Michael Boles Collection: https://dl.mospace.umsystem.edu//umkc/islandora/object/umkc:boles.

201 See the photographs in the University of Missouri–Kansas City's Robert Heishman Collection and the Michael Boles Collection.

202 For a good introduction to the topic, see L. Senelick, 'Queens of Clubs', in his *The Changing Room: Sex, Drag and Theatre* (New York, 2000), ch. 15.

203 E. Drorbaugh, 'Sliding Scales: Notes on Stormé DeLarverié and the Jewel Box Revue, the Cross-Dressed Woman on the Contemporary Stage, and the Invert', in L. Ferris (ed.), *Crossing the Stage: Controversies on Cross-Dressing* (New York, 1993), ch. 8. DeLarverié toured from 1955 to 1969 (p. 122).

204 B. Coleman, 'The Jewel Box Revue: America's Longest-Running, Touring Drag Show', *Theatre History Studies*, 17 (1997), 79–92, at 81.

205 See Internet Archive: https://archive.org/details/RaeBourbon-AroundTheWorldIn80Ways/09_Mr-Wong.mp3. For Bourbon, see D. Romesburg, 'Longevity and Limits in Rae Bourbon's Life in Motion', in T. T. Cotton (ed.), *Transgender Migrations: The Bodies, Borders, and Politics of Transition* (New York, 2012), ch. 7; and P. C. Byrne and D. W. Jackson, *Double Entendre: The Parallel Lives of Mae West and Rae Bourbon* (Albany, Ga., 2017).

206 Minette, *Recollections*.

207 Newton, *Mother Camp*, p. 112.

208 A. Willard, *Female Impersonation* (New York, 1971).

209 The songs and albums are on Internet Archive: https://archive.org/scarch.php?qucry=rae%20bourbon.

210 M. R. Gorman, *The Empress is a Man: Stories from the Life of José Sarria* (New York, 2013), pp. 144–5.

211 GLBT Historical Society, San Francisco (hereafter GLBTHS), James Dewsnap Papers and Artwork (#1999-19), Box 1 Interviews with Jackie Phillips 1994: Transcripts, Folder 2 Tapes 1–5.

212 Doyle, 'Club My-O-My'.

213 Newton, *Mother Camp*, p. 66.

214 GLBTHS, Interviews with Jackie Phillips 1994: Transcripts, Folder 3 Tapes 6–10.

215 Ibid.

216 Ibid.

217 Gorman, *Empress is a Man*, p. 145.

218 Minette, *Recollections*.

219 Doyle, 'Jimmy Callaway'.

220 J. J. Jeffreys, 'Who's No Lady? Excerpts from an Oral History of New York's 82 Club', *New York Folklore*, 19:1–2 (1993), 185–202, quote at 195.

221 Newton, *Mother Camp*, p. 10.

222 J. Rechy, *City of Night* (New York, 1984), p. 97. First published in 1963.

223 Ibid., pp. 101, 207.

224 Ibid., p. 151.

225 Ibid., pp. 101, 207, 284.

226 For example, the description in ibid., pp. 324–6.

227 Ibid., p. 117.

228 Ibid., p. 232.

229 Ibid., p. 116.

230 H. Selby, *Last Exit to Brooklyn* (New York, 1957), p. 48.

231 Ibid., p. 23.

232 Ibid., p. 45.

233 Ibid., p. 47.

234 Ibid., p. 71.

235 Ibid., p. 24.

236 Ibid., p. 215.

237 Ibid., pp. 30, 182, 218.

238 See album cover in Internet Archive: https://archive.org/details/RaeBourbon-LetMeTellYouAboutMyOperation/02_Oh-Doctor.mp3.

239 Paulson and Simpson, *Evening in the Garden of Allah*, pp. 38, 78, 122, 127.

240 Jeffreys, 'Oral History of New York's 82 Club', 191.

241 Newton, *Mother Camp*, pp. 12, 27, 102.

242 V. Prince, 'Change of Sex or Gender', *Transvestia*, 60 (1969), 53–65, quotes at 57, 59.

243 Both articles are in *Transvestia*, 60 (1969).

244 A. Stewart, 'Should I?????', *Transvestia*, 60 (1969), 74–8, quote at 74.

245 Beigel, 'A Week in Alice's Wonderland'.

246 K. Cummings, *Katherine's Diary: The Story of a Transsexual* (Tascott, NSW, 2008). First published in 1992.

247 Ibid., p. 156.

248 'Susanna Says', *Transvestia*, 61 (1970), 80–4, quote at 84.

249 See J. Vaccaro, 'A Transfeminist Archive of the Understanding Wife', in S. Stryker and others, 'Virtual Roundtable on *Transparent*', *Public Books*, 1 August 2015: www.publicbooks.org/artmedia/virtual-roundtable-on-transparent.

250 Benjamin, 'Transsexualism and Transvestism as Psycho-Somatic and Somatic-Psychic Syndromes', 220–1.

3 Blurring the Boundaries

1 D. Mason-Schrock, 'Transsexuals' Narrative Construction of the "True Self"', *Social Psychology Quarterly*, 59:3 (1996), 176–92, quote at 176. See, also, D. P. Schrock and L. L. Reid, 'Transsexuals' Sexual Stories', *Archives of Sexual Behavior*, 35:1 (2006), 75–86.

2 D. B. Hill, 'Dear Doctor Benjamin: Letters from Transsexual Youth (1963–1976)', *International Journal of Transgenderism*, 10:3–4 (2008), 149–70, quote at 167.

3 Ibid., 164.

4 GLBT Historical Society, San Francisco (hereafter GLBTHS), Oral History Collection 97-034, Transcript of Susan Stryker interview with Elliot Blackstone, 6 November 1996, p. 41.

5 J. M. Irvine, *Disorders of Desire: Sexuality and Gender in Modern American Sexology* (Philadelphia, 2005), p. 210.

6 S. Stryker, 'Perfect Day', in T. O'Keefe and K. Fox (eds.), *Trans People in Love* (New York, 2008), ch. 5, quote at pp. 46–7 (ellipsis in original).

7 S. Mesics, 'When Building a Better Vulva, Timing is Everything: A Personal Experience with the Evolution of MTF Genital Surgery', *TSQ*, 5:2 (2018), 245–50, quote at 246.

8 For example, D. Denny, 'Rachel and Me: A Commentary on

Gender: An Ethnomethodological Approach', *Feminism & Psychology*, 10:1 (2000), 62–5, quote at 63.

9 J. Meyerowitz, *How Sex Changed: A History of Transsexuality in the United States* (Cambridge, Mass., 2002), pp. 176, 187.

10 J. Morris, *Conundrum* (New York, 1986), p. 3. First published in 1974.

11 M. Martino, *Emergence: A Transsexual Autobiography* (New York, 1977), p. xii.

12 Ibid., p. 3.

13 Ibid., pp. 13, 17.

14 B. L. Hausman, *Changing Sex: Transsexualism, Technology, and the Idea of Gender* (Durham, NC, 1995), p. 143.

15 T. Kando, *Sex Change: The Achievement of Gender Identity Among Feminized Transsexuals* (Springfield, Ill., 1973), p. 145.

16 Ibid., p. 24.

17 Morris, *Conundrum*, pp. 152–3, 158.

18 I. B. Pauly, 'Adult Manifestations of Male Transsexualism', in R. Green and J. Money (eds.), *Transsexualism and Sex Reassignment* (Baltimore, 1969), ch. 3, quote at p. 38.

19 Ibid., pp. 53–6.

20 I. B. Pauly, 'Adult Manifestations of Female Transsexualism', in Green and Money (eds.), *Transsexualism and Sex Reassignment*, ch. 4, quote at p. 60.

21 J. Money and C. Primrose, 'Sexual Dimorphism and Dissociation in the Psychology of Male Transsexuals', in Green and Money (eds.), *Transsexualism and Sex Reassignment*, ch. 6, quote at p. 127.

22 R. J. Stoller, 'Parental Influences in Male Transsexualism', in Green and Money (eds.), *Transsexualism and Sex Reassignment*, ch. 8, quote at p. 166.

23 Ibid.

24 Money and Primrose, 'Sexual Dimorphism and Dissociation', p. 116.

25 Ibid., p. 119.

26 Ibid., p. 121.

27 Ibid., p. 124.

28 J. Money and J. G. Brennan, 'Sexual Dimorphism in the Psychology

of Female Transsexuals', in Green and Money (eds.), *Transsexualism and Sex Reassignment*, ch. 7.

29 Ibid., p. 139.

30 Ibid., pp. 140–1.

31 Ibid., p. 142.

32 Ibid., p. 141.

33 Ibid., pp. 144–5.

34 H. Benjamin, 'Trans-Sexualism and Transvestism', in David O. Cauldwell (ed.), *Transvestism . . . Men in Female Dress* (New York, 1956), ch. 1, quotes at p. 18.

35 H. Benjamin, *The Transsexual Phenomenon* (New York, 1966), pp. 21–2.

36 Kinsey Institute, University of Indiana, Bloomington (hereafter KI), The Harry Benjamin Collection (hereafter Benjamin), Box 3, Series 2c, Correspondence, Folder: Belt, Dr Elmer, 1958–9, Letter: 2 June 1958.

37 Ibid., Letter: 28 July 1958.

38 Ibid., Letter: 28 Nov. 1958.

39 Benjamin, Box 14, Series 3a, Clinical/Research: Patient Files, Folder 16: Meetings-Male TS Groups, 16 Feb. 1969.

40 Ibid., 12 Jan. 1969.

41 Benjamin, Meetings-Female TS Groups, 16 June 1968.

42 Benjamin, Meetings-Male TS Groups, 16 Feb. 1969.

43 Money and Brennan, 'Sexual Dimorphism in the Psychology of Female Transsexuals', p. 146.

44 Benjamin, *Transsexual Phenomenon*, pp. 232–47.

45 Ibid., pp. 245–6.

46 Ibid., p. 235.

47 Ibid., p. 236.

48 Ibid., pp. 237–47.

49 Ibid., pp. 240, 241.

50 L. M. Lothstein, *Female-to-Male Transsexualism: Historical, Clinical and Theoretical Issues* (Boston, Mass., 1983), p. 86.

51 Ibid., pp. 95–6.

52 Ibid., pp. 101–2.

53 Ibid., p. 119.

54 Ibid., p. 131.

55 Ibid., p. 139.
56 Ibid., p. 141.
57 Ibid., p. 131.
58 Ibid., p. 109.
59 Ibid., p. 131.
60 Ibid., pp. 134–5.
61 Ibid., p. 273.
62 Ibid., p. 274.
63 Ibid.
64 Ibid.
65 E. L. Kennedy and M. Davis, 'The Reproduction of Butch–Fem Roles: A Social Constructionist Approach', in K. Peiss and C. Simmons (eds.), *Passion and Power: Sexuality in History* (Philadelphia, 1989), ch. 13. They were writing about fem roles, too.
66 Ibid., p. 249.
67 J. Green, *Becoming a Visible Man* (Nashville, 2004), p. 17.
68 Ibid., p. 56.
69 H. Devor, *FTM: Female-to-Male Transsexuals in Society* (Bloomington and Indianapolis, 1999), pp. 344–5.
70 H. Rubin, *Self-Made Men: Identity and Embodiment Among Transsexual Men* (Nashville, 2003), p. 7.
71 Ibid., ch. 2.
72 Meyerowitz, *How Sex Changed*, pp. 171–3; Irvine, *Disorders of Desire*, pp. 214–15.
73 GLBTHS, San Francisco LGBT Groups Ephemera Collection (#GRP EPH), Groups Box, Folder: Erickson Educational Foundation, *Counselling the Transexual [sic]: Five Conversations with Professionals in Transexual [sic] Therapy* (Baton Rouge, La., c.1973), pp. 8–9, 26.
74 V. LeMans, *Take My Tool* (Los Angeles, 1968), quote at p. 121.
75 A. J. Morgan, 'Psychotherapy for Transsexual Candidates Screened Out of Surgery', *Archives of Sexual Behavior*, 7:4 (1978), 273–83, quote at 277.
76 See H. Apple, 'The $10,000 Woman: Trans Artifacts in the Pittsburgh Queer History Project Archive', *TSQ*, 2:4 (2015), 553–64.

77 J. County and R. Smith, *Man Enough to be a Woman* (New York, 1995), p. 30.

78 J. Gamson, *The Fabulous Sylvester* (New York, 2005), pp. 1–10, quote at p. 5.

79 *The Queen* (1968: dir. Frank Simon).

80 J. Corr, 'Different, Yet the Same', *The Philadelphia Inquirer*, 18 April 1989: http://articles.philly.com/1989-04-18/news/26144375_1_gay-man-gay-men-transsexual.

81 *Queens at Heart* (1967: dir. unknown).

82 *The Queen* (1968: dir. Frank Simon).

83 S. Rivera and M. P. Johnson, *Street Transvestite Action Revolutionaries: Survival, Revolt, and Queer Antagonist Struggle* (Untorelli Press [no place of publication], 2013), pp. 47–8.

84 Members of the Gay and Lesbian Historical Society, 'MTF Transgender Activism in the Tenderloin and Beyond, 1966–1975: Commentary and Interview with Elliot Blackstone', *GLQ*, 4:2 (1998), 349–72, quote at 354.

85 J. P. Driscoll, 'Transsexuals', *Transaction*, 8:5–6 (1971), 28–68, quote at 32.

86 For a little more detail, see the study on which Driscoll's article is based: J. P. Driscoll, 'The Transsexuals', San Francisco State College MA, 1969, pp. 41–2.

87 Driscoll, 'The Transsexuals', pp. 45–7.

88 Ibid., p. 35.

89 Members of the Gay and Lesbian Historical Society, 'MTF Transgender Activism in the Tenderloin', p. 360.

90 M. Foerster, 'On the History of Transsexuals in France', in C. Zabus and D. Coad (eds.), *Transgender Experience: Place, Ethnicity, and Visibility* (New York, 2014), ch. 1.

91 M. A. Costa, *Reverse Sex* (London, 1965), p. 92.

92 A. Ashley and D. Thompson, *The First Lady* (London, 2006), ch. 18.

93 Ibid., p. 85.

94 Ibid., pp. 86–7, 96.

95 See ibid., index: Lear, Amanda. See, also, A. Lear, *Persistence of Memory: A Personal Biography of Salvador Dali* (Bethesda, Md., 1987).

96 C. Strömholm, *Les Amies de Place Blanche* (Paris, 2011). First published in 1983.

97 Members of the Gay and Lesbian Historical Society, 'MTF Transgender Activism in the Tenderloin', p. 360; GLBTHS, Oral History Collection 97-034, Transcript of Susan Stryker interview with Elliot Blackstone, 6 November 1996, p. 24.

98 Driscoll, 'Transsexuals', p. 29; Driscoll, 'The Transsexuals', p. 69.

99 Driscoll, 'Transsexuals', p. 35.

100 S/B Glick, 'Down on the Tenderloin', *Good Times*, 21 August 1970.

101 Members of the Gay and Lesbian Historical Society, 'MTF Transgender Activism in the Tenderloin', p. 369, n. 8.

102 B. L. Hillman, '"The Most Profoundly Revolutionary Act a Homosexual Can Engage In": Drag and the Politics of Gender Presentation in the San Francisco Gay Liberation Movement, 1964–1972', *Journal of the History of Sexuality*, 20:1 (2011), 153–81, quote at 165.

103 J. E. Siegel and P. Feinberg, *Another Washington* (Washington, DC, 1976). Siegel wrote the text and Feinberg took the photographs.

104 Ibid., p. 28.

105 Ibid., pp. 30–3.

106 Ibid., pp. 37–9.

107 Ibid., pp. 41–3.

108 S. Brecht, *Queer Theatre* (New York, 1986), p. 54 (the original essay was written in 1968).

109 S. Sontag, 'Jack Smith's *Flaming Creatures*', in her *Against Interpretation* (New York, 1996), pp. 226–31, quote at p. 230. For Smith and *Flaming Creatures*, see E. Leffingwell, C. Kismaric, and M. Heiferman (eds.), *Jack Smith: Flaming Creature* (Long Island City, NY, 1997); J. Hoberman, *On Jack Smith's Flaming Creatures* (New York, 2001); D. Johnson, *Glorious Catastrophe: Jack Smith, Performance and Visual Culture* (Manchester, 2012), esp. ch. 3.

110 R. Tavel, 'The Banana Diary: The Story of Andy Warhol's "Harlot"', in M. O'Pray (ed.), *Andy Warhol Film Factory* (London, 1989), pp. 66–93, quote at p. 85. The essay was written in 1965.

111 Ibid.

112 Ibid., p. 68.

113 See D. Crimp, 'Our Kind of Movie': The Films of Andy Warhol (Cambridge, Mass., 2012), Index: 'Montez, Mario'.

114 Vogue, 1 June 1972; M. Yacowar, The Films of Paul Morrissey (Cambridge, 1993), p. 44.

115 C. Darling and J. Rasin, Candy Darling: Memoirs of an Andy Warhol Superstar (New York, 2015), Ebook, loc. 740.

116 Ibid., loc. 614.

117 Ibid., loc. 465.

118 Ibid., loc. 201.

119 Ibid., loc. 558.

120 Ibid., loc. 215.

121 Ibid., loc. 1536.

122 R. Avedon, 'Andy Warhol and Members of the Factory, 30 October 1969': www.tate.org.uk/art/artworks/avedon-andy-warhol-and-members-of-the-factory-30-october-1969-p13101.

123 H. Woodlawn and J. Copeland, A Low Life in High Heels: The Holly Woodlawn Story (New York, 1991).

124 Ibid., p. 73.

125 Ibid., pp. 92–3.

126 Ibid., p. 94.

127 Ibid., p. 96.

128 Ibid., pp. 110–11.

129 Ibid., pp. 98–102.

130 Ibid., p. 278.

131 Ibid., p. 76.

132 G. Palladini, 'Queer Kinship in the New York Underground: On the "Life and Legend" of Jackie Curtis', Contemporary Theatre Review, 21:2 (2011), 126–53.

133 See the images in ibid. See, also, C. B. Highberger, Superstar in a Housedress: The Life and Legend of Jackie Curtis (New York, 2005), DVD included.

134 Palladini, 'Queer Kinship', 136.

135 Woodlawn and Copeland, Low Life in High Heels, p. 77.

136 A. Warhol and P. Hackett, Popism: The Warhol Sixties (New York, 1990), p. 224; Woodlawn and Copeland, Low Life in High Heels, pp. 76–7.

137 R. Regelson, 'Not a Boy, Not a Girl, Just Me', *New York Times*, 2 November 1969.

138 Ibid.

139 Ibid.

140 Warhol and Hackett, *Popism*, p. 293.

141 Tavel, 'Banana Diary', pp. 66, 70.

142 Ibid., p. 85.

143 H. Oiticia, 'Héliotape with Mario Montez (1971)', *Criticism*, 56:2 (2014), 379–404, quotes at 396.

144 D. Crimp, 'Mario Montez, For Shame', in his *Our Kind of Movie*, pp. 2–36, quotes at pp. 22, 26–7.

145 Warhol and Hackett, *Popism*, p. 223.

146 B. Colacello, *Holy Terror: Andy Warhol Close Up* (New York, 2000), pp. 221–2. First published in 1990. Also partially quoted in P. Hickson, 'Warhol and Mapplethorpe: Guise and Dolls', in P. Hickson (ed.), *Warhol and Mapplethorpe: Guise and Dolls* (New Haven, 2015), pp. 1–19, quote at p. 6.

147 For the portraits, see N. Printz and S. King-Nero (eds.), *The Andy Warhol Catalogue Raisonné*, Vol. IV: *Paintings and Sculpture, Late 1974–1976* (New York, 2014), pp. 22–203, 548–62.

148 J. Flatley, *Like Andy Warhol* (Chicago, 2017), p. 221.

149 For the Cockettes, apart from later references, see M. Gaines, 'The Cockettes, Sylvester, and Performance as Life', in his *Black Performance on the Outskirts of the Left: A History of the Impossible* (New York, 2017), ch. 4. Although its focus is on Sylvester, Gamson's *The Fabulous Sylvester* contains probably the best account of the Cockettes.

150 M. Henry, 'Cock-Ettes', *Good Times*, 21 August 1970.

151 The New York Public Library for the Performing Arts (hereafter NYPLPA), Billy Rose Theatre Division, Martin Worman Papers MSS 2009–002 (hereafter Worman Papers), Box 19, Folder 6: Martin Worman, 'The Cockettes: Their First Year and How I Became One', Term Paper 1987, Preface.

152 Ibid., Box 19, Folder 7: Manuscripts of Thesis Material, 'Thesis Proposal', 3–4.

153 J. Le Zotte, *From Goodwill to Grunge: A History of Secondhand Styles and Alternative Economies* (Chapel Hill, NC, 2017), pp. 206–7.

154 GLBTHS, #GRP EPH, Groups Box, Folder: Cockettes: *Pristine Condition's Tri-Sexual Bicentennial Universal Calendar for 1976.*

155 J. Bryan-Wilson, *Fray: Art and Textile Politics* (Chicago, 2017), pp. 42–3. Emphasis in original.

156 Ibid., pp. 42–72. See, also, Le Zotte, *From Goodwill to Grunge*, ch. 6: 'Genderfuck and the Boyfriend Look'.

157 M. Gussow, 'Theater: Coast Transvestite Troup', *New York Times*, 9 Nov. 1971.

158 R. Reed, 'Better a Tinsel Queen than a Golden Toad', *Chicago Tribune*, 19 Sept. 1971.

159 Worman Papers, Box 19, Folder 7: Manuscripts of Thesis Material, 'Thesis Proposal', 4.

160 Reed, 'Better a Tinsel Queen'.

161 Worman Papers, Box 18, Folder 5: Worman's Research Files, 1970–1993, Cockettes' Interview Transcripts, 'A Dish before Dying: Sylvester's Last Interview', 2.

162 Worman Papers, Box 17, Folder 10: Worman's Research Files, 1970–1993, Cockettes' Interview Transcripts, Dolores Deluxe, 30 August 1987, 4.

163 B. Lake and A. Orloff, *The Unsinkable Bambi Lake* (San Francisco, 2017), p. 148.

164 Ibid., pp. 70, 148–9.

165 J. Camicia, *My Dear Sweet Self: A Hot Peach Life* (Silverton, Oreg., 2013). For Johnson and Ross in *Ladies and Gentlemen*, see Printz and King-Nero (eds.), *Andy Warhol Catalogue Raisonné*, Vol. IV, pp. 41–4, 46–51, 52–5, 70–4, 87–101, 168–73, 199–201, 553, 556, 561, 562–3.

166 C. Ludlam, *Ridiculous Theatre: Scourge of Human Folly – The Essays and Opinions of Charles Ludlam*, ed. S. Samuels (New York, 1992), p. 22.

167 Ibid., p. 153.

168 Ibid., pp. 41, 43.

169 Ibid., p. 25. For the play, see C. Ludlum, *The Complete Plays of Charles Ludlum* (New York, 1989), pp. 115–41, with a photograph of Montez on p. 116.

170 H. N. R. Ramírez, 'Sharing Queer Authorities: Collaborating for Transgender Latina and Gay Latino Historical Meanings', in N. A.

Boyd and H. N. R. Ramírez (eds.), *Bodies of Evidence: The Practice of Queer Oral History* (New York, 2012), ch. 10, quote at p. 186.

171 Ibid., p. 189.

172 Ibid., p. 192.

173 Ibid.

174 For STAR, see S. L. Cohen, 'Street Transvestite Action Revolutionaries (S.T.A.R.)', in his *The Gay Liberation Youth Movement in New York: 'An Army of Lovers Cannot Fail'* (New York, 2008), ch. 5.

175 A. Young, 'Rapping with a Street Transvestite Revolutionary: An Interview with Marcia Johnson', in K. Jay and A. Young (eds.), *Out of the Closets: Voices of Gay Liberation* (New York, 1972), pp. 112–20. See, also, Rivera and Johnson, *Street Transvestite Action Revolutionaries*. Whitney Strub begins his recent survey of the history of gay liberation with Johnson and STAR: W. Strub, 'Gay Liberation (1963–1980)', in D. Romesburg (ed.), *The Routledge History of Queer America* (New York, 2018), ch. 6.

176 H. Benjamin, 'Transvestism and Transsexualism in the Male and Female', *The Journal of Sex Research*, 3:2 (1967), 107–27. He treated 340 male-to-female transvestites, 242 of whom were classified as transsexual and 76 of whom had surgery.

177 C. L. Ihlenfeld, 'Thoughts on the Treatment of Transsexuals', *Journal of Contemporary Psychotherapy*, 6:1 (1973), 63–9, quote at 69.

178 Benjamin, 'Transvestism and Transsexualism', 108.

179 Kando, *Sex Change*, pp. 4, 127–9.

180 Ibid., pp. 127–8.

181 GLBTHS, Oral History Collection 07-002, Transcript of Susan Stryker interview with Major, 29 January 1998, pp. 1, 5.

182 GLBTHS, Oral History Collection 97-036, Transcript of Susan Stryker interview with Regina Elizabeth McQueen, 17 July 1997, p. 11.

183 Ibid., p. 9.

184 Ibid., p. 13.

185 Ibid., p. 15.

186 Ibid., p. 13.

187 GLBTHS, Oral History Collection 97-040, Transcript of Susan

Stryker interview with Elishia [Aleshia] Brevard Crenshaw, 2 August 1997, p. 54.

188 A. Brevard, *The Woman I Was Not Born to Be: A Transsexual Journey* (Philadelphia, 2001), p. 44.

189 Transcript of Susan Stryker interview with Elishia [Aleshia] Brevard Crenshaw, 2 August 1997, p. 9.

190 Brevard, *The Woman I Was Not Born to Be*, p. 49.

191 Ibid., pp. 50, 61.

192 Ibid., p. 77.

193 L. Siegel and A. Zitrin, 'Transsexuals in the New York Welfare Population: The Function of Illusion in Transsexuality', *Archives of Sexual Behavior*, 7:4 (1978), 285–90.

194 Ibid., 286.

195 Ibid., 287.

196 Ibid.

197 S. Cooke, 'Trannywood', *TNT: Transsexual News Telegraph*, 9 (2000).

198 Ibid. For identification, see www.jiscmail.ac.uk/cgi-bin/webad min?A2=trans-theory;2ab3efa4.0010.

199 GLBTHS, Oral History Collection 97-039, Transcript of Susan Stryker interview with Suzan Cooke, 10 January 1998, p. 60.

200 GLBTHS, #GRP EPH, Groups Box, Folder: Golden Gate Girls / Golden Gate Guys Membership Directory, *c.*1979.

201 GLBTHS Periodical Collection: *Drag: A Magazine About the Transvestite*, 7:25 (1977).

202 N. Goldin, *The Other Side* (Zurich, 1995), p. 5.

203 See A. Badertscher, *Baltimore Portraits* (Durham, NC, 1999). The Badertscher archive is divided between his residence in Baltimore (hereafter BB) and the Leslie–Lohman Museum of Gay and Lesbian Art in New York (hereafter L–L). The figures are from high-quality digital images of the original photographs in Badertscher's personal archive.

204 See B. Bogdan and B. Reay, 'Queer Baltimore: The Photography of Amos Badertscher', in B. Reay, *Sex in the Archives: Writing American Sexual Histories* (Manchester, 2019), ch. 9.

205 Ibid., pp. 266–8.

206 A. Badertscher, 'Biographies of Models', BB.

207 'Todd', 1975, in Badertscher, *Baltimore Portraits*, unpaginated plates.

208 'Frank', 1974, L–L.

209 'Miss Natalie', 1975, L–L.

210 T. Curtain, 'A Baltimore Essay: Photography, Sexuality, Community', in Badertscher, *Baltimore Portraits*, pp. 1–11, quote at p. 1.

211 J. Money [moderator], 'Open Forum', *Archives of Sexual Behavior*, 7:4 (1978), 387–415, quote at 391.

212 Archives & Special Collections, Augustus C. Long Health Sciences Library, Columbia University Medical Center, New York, Ethel Spector Person Papers (hereafter Person), Box 14, Folder 6: Transsexualism Case Histories: Male-to-Female, Copy of an assessment of B. S., 1 August 1971.

213 R. J. Stoller, *The Transsexual Experiment* (London, 1975), p. 153.

214 R. J. Stoller, *Sex and Gender: On the Development of Masculinity and Femininity* (New York, 1968), pp. 188–9.

215 Department of Special Collections, Charles E. Young Research Library, UCLA, Robert J. Stoller Papers, Box 3: Transvestism/ Transsexualism, L. E. to R. Stoller, 21 May 1973.

216 Stoller, *Transsexual Experiment*, p. 153.

217 Ibid., pp. 153–4.

218 Ibid., pp. 179–80.

219 Ibid., p. 177.

220 Ibid., p. 158.

221 Ibid., pp. 158, 178.

222 Person, Box 14, Folder 5: Transsexualism Case History: Male-to-Female, W., n.d., c.1970s.

223 D. Greene, *Photographs* (Chicago, 2012).

224 C. Lonc, 'Genderfuck and Its Delights', in W. Leyland (ed.), *Gay Roots: Twenty Years of Gay Sunshine* (San Francisco, 1991), pp. 223–6, quote at p. 225. Originally published in 1974.

225 L. Humphreys, *Out of the Closets: The Sociology of Homosexual Liberation* (Englewood Cliffs, 1972), p. 164.

226 B. L. Hillman, '"The Most Profoundly Revolutionary Act a Homosexual Can Engage In": Drag and the Politics of Gender Presentation in the San Francisco Gay Liberation Movement,

1964–1972', *Journal of the History of Sexuality*, 20:1 (2011), 153–81. See, also, B. L. Hillman, *Dressing for the Culture Wars: Style and the Politics of Self-Presentation in the 1960s and 1970s* (Lincoln, Nebr., 2015), ch. 4.

227 Hillman, 'Most Profoundly Revolutionary Act', 172.

228 R. Meyer, '*Gay Power* Circa 1970: Visual Strategies for Sexual Revolution', *GLQ*, 12:3 (2006), 441–64, at 434 and 436.

229 County and Smith, *Man Enough to be a Woman*, quotes at pp. 21, 139.

230 Ibid., p. 30.

231 Ibid., p. 176, for County describing herself as a transsexual just before she says (on p. 177) that she would rather have a facelift than gender-confirming surgery.

4 Backlash

1 L. S. Kubie, 'The Drive to Become Both Sexes', *The Psychoanalytic Quarterly*, 43 (1974), 349–426, quote at 382, n. 5.

2 H. Benjamin, *The Transsexual Phenomenon* (New York, 1966); R. J. Stoller, *Sex and Gender: On the Development of Masculinity and Femininity* (New York, 1968); R. Green and J. Money (eds.), *Transsexualism and Sex Reassignment* (Baltimore, 1969). See, also, R. J. Stoller, *The Transsexual Experiment* (London, 1975).

3 Essentially, 5 out of more than 300 pages: J. Meyerowitz, *How Sex Changed: A History of Transsexuality in the United States* (Cambridge, Mass., 2002), pp. 266–70, quote at p. 254.

4 S. Stryker, *Transgender History* (Berkeley, 2008).

5 J. A. M. Meerloo, 'Change of Sex and Collaboration with the Psychosis', *The American Journal of Psychiatry*, 124:2 (1967), 167–8, quote at 167.

6 L. S. Kubie and J. B. Mackie, 'Critical Issues Raised by Operations for Gender Transmutation', *The Journal of Nervous and Mental Disease*, 147:5 (1968), 431–43, quotes at 431.

7 Ibid., 431.

8 Ibid., 435.

9 Ibid.

10 Ibid.

11 Ibid., 436.

12 Stoller, *Transsexual Experiment*, ch. 19, quote at p. 247.

13 Ibid., p. 248.

14 Ibid.

15 Ibid., p. 254.

16 Ibid., p. 255.

17 C. Conn, *Canary: The Story of a Transsexual* (Los Angeles, 1974), p. 280.

18 Ibid., p. 284.

19 E. Rutherford, *Nine Lives: The Autobiography of Erica Rutherford* (Charlottetown, PE, 1993), p. 196.

20 R. Bogdan (ed.), *Being Different: The Autobiography of Jane Fry* (New York, 1974), pp. 135–6.

21 Ibid.

22 Ibid., p. 20.

23 Ibid.

24 Ibid., p. 27.

25 Ibid.

26 R. Green, R. J. Stoller, and C. MacAndrew, 'Attitudes Toward Sex Transformation Procedures', *Archives of General Psychiatry*, 15 (1966), 178–82.

27 C. W. Socarides, 'The Desire for Sexual Transformation: A Psychiatric Evaluation of Transsexualism', *The American Journal of Psychiatry*, 125:10 (1969), 1419–25.

28 E. Sagarin, *Odd Man In: Societies of Deviants in America* (Chicago, 1969), p. 122. For Sagarin/Corey, see M. Duberman, 'The "Father" of the Homophile Movement', in his *Left Out: The Politics of Exclusion: Essays 1964–2002* (Cambridge, Mass., 2002), pp. 59–94.

29 Sagarin, *Odd Man In*, p. 117.

30 Ibid., p. 130.

31 E. Sagarin, 'Review of *Being Different: The Autobiography of Jane Fry* by Robert Bogdan', *The Journal of Sex Research*, 11:2 (1975), 163–4, quote at 163.

32 E. Sagarin, 'Ideology as a Factor in the Consideration of Deviance', *The Journal of Sex Research*, 4:2 (1968), 84–94, quotes at 86 and 88.

33 H. Garfinkel, 'Passing and the Managed Achievement of Sex Status in an "Intersexed" Person Part 1', in his *Studies in Ethnomethodology*

(Englewood Cliffs, NJ, 1967), ch. 5. See also, 'Appendix to Chapter 5' in the same volume, pp. 285–8.

34 Ibid., p. 119.
35 Ibid., pp. 119, 126.
36 Ibid., pp. 120, 285.
37 Ibid., p. 119, n. 1.
38 Ibid., p. 288.
39 K. Schilt, 'The Importance of Being Agnes', *Symbolic Interaction*, 39:2 (2016), 287–94, quote at 292.
40 Garfinkel, *Studies in Ethnomethodology*, p. 174.
41 Ibid., p. 164.
42 Ibid., p. 165.
43 Ibid., p. 147.
44 Ibid., p. 128.
45 Ibid., p. 178.
46 Ibid., p. 163.
47 Kinsey Institute, University of Indiana, Bloomington (hereafter KI), The Harry Benjamin Collection (hereafter Benjamin), Box 5, Series 2c, Correspondence, Folder: Belt, Dr Elmer, 1965–71, Letter: 24 March 1969.
48 Ibid.
49 Stoller, *Transsexual Experiment*, ch. 7.
50 J. Money and J. G. Brennan, 'Heterosexual Vs. Homosexual Attitudes: Male Partners' Perception of the Feminine Image of Male Transsexuals', *The Journal of Sex Research*, 6:3 (1970), 193–209, quotes at 201, 202.
51 Ibid., 207.
52 L. M. Lothstein, *Female-to-Male Transsexualism: Historical, Clinical and Theoretical Issues* (Boston, Mass., 1983), p. 233.
53 Ibid., p. 9.
54 Ibid., p. 281.
55 Ibid., pp. 277–9.
56 Ibid., p. 279.
57 S. R. Wolf and others, 'Psychiatric Aspects of Transsexual Surgery Management', *The Journal of Nervous and Mental Disease*, 147:5 (1968), 525–31, quotes at 525.
58 J. P. Driscoll 'Author's Reply', in *Transaction*, 8:12 (1971), 12.

59 M. A. Costa, *Reverse Sex* (London, 1965), p. 117.

60 Conn, *Canary*, p. 310.

61 N. Hunt, *Mirror Image* (New York, 1978), pp. 205, 210.

62 M. Martino, *Emergence: A Transsexual Autobiography* (New York, 1977), p. 262.

63 A. Brevard, *The Woman I Was Not Born to Be: A Transsexual Journey* (Philadelphia, 2001), p. 84.

64 Ibid., p. 83.

65 P. Morgan, *The Man-Maid Doll* (Secaucus, NJ, 1973), p. 60.

66 R. Richards with J. Ames, *Second Serve* (New York, 1984), p. 282. First published in 1983.

67 Ibid., p. 281.

68 Garfinkel, *Studies in Ethnomethodology*, p. 153.

69 H. W. Jones, 'Operative Treatment of the Male Transsexual', in Green and Money (eds.), *Transsexualism and Sex Reassignment*, ch. 22.

70 P. H. Thompson, 'Apparatus to Maintain Adequate Vaginal Size Postoperatively in the Male Transsexual Patient', in Green and Money (eds.), *Transsexualism and Sex Reassignment*, ch. 23. See, also, L. Wollman, 'Office Management of the Postoperative Male Transsexual', in Green and Money (eds.), *Transsexualism and Sex Reassignment*, ch. 24.

71 Jones, 'Operative Treatment of the Male Transsexual', p. 314.

72 For example, D. R. Laub and P. Gandy (eds.), *Proceedings of the Second Interdisciplinary Symposium on Gender Dysphoria Syndrome* (Stanford, 1973), pp. 137, 138, 146, 150, 151, 160–2, 166–7, 180, 184–5.

73 J. E. Hoopes, 'Operative Treatment of the Female Transsexual', in Green and Money (eds.), *Transsexualism and Sex Reassignment*, ch. 25, quotes at pp. 336, 337, 339.

74 Ibid., p. 339.

75 Ibid., pp. 341–2.

76 See Figure 6 in ibid., p. 352.

77 J. Green, *Becoming a Visible Man* (Nashville, 2004), p. 107.

78 Bogdan (ed.), *Being Different*, p. 105.

79 Ibid., p. 214.

80 Ibid.

81 Ibid., p. 216.

82 E. L. Weitzman, C. A. Shamoian, and N. Golosow, 'Identity Diffusion and the Transsexual Resolution', *The Journal of Nervous and Mental Disease*, 151:5 (1970), 295–302, quote at 297.

83 Ibid., 296.

84 Ibid., 295.

85 L. M. Lothstein, 'Countertransference Reactions to Gender Dysphoric Patients: Implications for Psychotherapy', *Psychotherapy: Theory, Research and Practice*, 14:1 (1977), 21–31, esp. 22.

86 Ibid., 21, 24.

87 Ibid., 25, 27.

88 Ibid., 29.

89 Ibid., 26.

90 Ibid., 27.

91 Ibid., 27, 28.

92 Lothstein, *Female-to-Male Transsexualism*, pp. 82–3.

93 Ibid., p. 68.

94 Ibid., pp. 71, 72.

95 Ibid., p. 71.

96 Ibid., pp. 72, 74.

97 Ibid., p. 147.

98 Ibid., p. 155.

99 Ibid.

100 R. Green, L. E. Newman, and R. J. Stoller, 'Treatment of Boyhood "Transsexualism": An Interim Report of Four Years' Experience', *Archives of General Psychiatry*, 26 (1972), 213–17.

101 See R. Green and J. Money, 'Incongruous Gender Role: Nongenital Manifestations in Prepubertal Boys', *The Journal of Nervous and Mental Disease*, 131:2 (1960), 160–8.

102 R. Green and J. Money, 'Effeminacy in Prepubertal Boys: Summary of Eleven Cases and Recommendation for Case Management', *Pediatrics*, 27 (1961), 286–91.

103 Green, Newman, and Stoller, 'Treatment of Boyhood "Transsexualism"', 213.

104 G. A. Rekers, O. I. Lovaas, and B. Low, 'The Behavioral Treatment of a "Transsexual" Preadolescent Boy', *Journal of Abnormal Child Psychology*, 2:2 (1974), 99–116, quote at 99.

105 G. A. Rekers and O. I. Lovaas, 'Behavioral Treatment of Deviant Sex-Role Behaviors in a Male Child', *Journal of Applied Behavioural Analysis*, 7:2 (1974), 173–90, quote at 188.

106 Green, Newman, and Stoller, 'Treatment of Boyhood "Transsexualism"', 213.

107 Rekers, Lovaas, and Low, 'Behavioral Treatment', 102.

108 Ibid., 113.

109 K. E. Bryant, 'The Politics of Pathology and the Making of Gender Identity Disorder', University of California Santa Barbara Ph.D., 2007, p. 4.

110 R. Green, *Sexual Identity Conflict in Children and Adults* (New York, 1974), p. 188.

111 R. Green, *The Sissy Boy Syndrome and the Development of Homosexuality* (New Haven, 1987), p. 226.

112 Rekers, Lovaas, and Low, 'Behavioral Treatment', 114.

113 Ibid., 115. Bryant has discussed Rekers's work, and the case of Carl, in far more detail: Bryant, 'Politics of Pathology and the Making of Gender Identity Disorder', pp. 128–49.

114 Green, Newman, and Stoller, 'Treatment of Boyhood "Transsexualism"', 214.

115 Stoller, *Transsexual Experiment*, pp. 102, 104, 105 (for quote).

116 Green, *The Sissy Boy Syndrome*, p. 271.

117 Examples from Rekers and Lovaas, 'Behavioral Treatment of Deviant Sex-Role Behaviors in a Male Child', p. 186; Green, *The Sissy Boy Syndrome*, pp. 119, 136–7, 149, 194–5, 267.

118 Rekers and Lovaas, 'Behavioral Treatment of Deviant Sex-Role Behaviors in a Male Child', 176.

119 Bryant, 'Politics of Pathology and the Making of Gender Identity Disorder', pp. 141–9.

120 R. R. Greenson, 'A Transvestite Boy and a Hypothesis', *International Journal of Psycho-Analysis*, 47 (1966), 396–403, quote at 396.

121 Green, Newman, and Stoller, 'Treatment of Boyhood "Transsexualism"', 213.

122 P. Burke, *Gender Shock: Exploding the Myths of Male and Female* (New York, 1996).

123 Green, Newman, and Stoller, 'Treatment of Boyhood "Transsexualism"', 217.

124 Green, *Sexual Identity Conflict*, p. 308.

125 Stoller, *Transsexual Experiment*, p. 37.

126 R. J. Stoller, 'Transvestites' Women', *The American Journal of Psychiatry*, 124:3 (1967), 333–9, quote at 333.

127 Ibid., 333–4.

128 Stoller, *Transsexual Experiment*, p. 42.

129 Stoller, 'Transvestites' Women', 333–4.

130 Ibid., 338.

131 J. Money, *Gay, Straight, and In-Between: The Sexology of Erotic Orientation* (New York, 1988), p. 83. He was reporting a study published in 1984.

132 Ibid., p. 82.

133 Ibid.

134 R. J. Stoller, 'Etiological Factors in Female Transsexualism: A First Approximation', *Archives of Sexual Behavior*, 2:1 (1972), 47–64, quote at 47.

135 J. Money, 'Use of an Androgen-Depleting Hormone in the Treatment of Male Sex Offenders', *The Journal of Sex Research*, 6:3 (1970), 165–72; Money, *Gay, Straight, and In-Between*, p. 97.

136 Money, *Gay, Straight, and In-Between*, pp. 232–7.

137 Ibid., p. 97.

138 Richards, *Second Serve*, p. 210.

139 A. Jagose, 'Behaviorism's Queer Trace: Sexuality and Orgasmic Reconditioning', in her *Orgasmology* (Durham, NC, 2013), ch. 3. For a recent oral history of some of the (now elderly or deceased) nurses and patients involved in the treatment of homosexuality and transvestism by aversion therapy (including homosexual nurses who treated homosexuality), see T. Dickinson, *'Curing Queers': Mental Nurses and their Patients, 1935–74* (Manchester, 2015).

140 N. I. Lavin and others, 'Behavior Therapy in a Case of Transvestism', *The Journal of Nervous and Mental Disease*, 133 (1961), 346–53, quote at 346.

141 Ibid.

142 Ibid., 347.

143 Ibid., 349.

144 J. C. Barker and others, 'Behaviour Therapy in a Case of Transvestism', *The Lancet*, 277:7175 (1961), 510.

145 C. B. Blakemore and others, 'The Application of Faradic Aversion Conditioning in a Case of Transvestism', *Behaviour Research and Therapy*, 1:1 (1963), 29–34.

146 Lavin and others, 'Behavior Therapy in a Case of Transvestism', 346.

147 M. G. Gelder and I. M. Marks, 'Aversion Treatment in Transvestism and Transsexualism', in Green and Money (eds.), *Transsexualism and Sex Reassignment*, ch. 27, quotes at p. 394.

148 Ibid., p. 403.

149 D. H. Barlow, E. J. Reynolds, and W. S. Agras, 'Gender Identity Change in a Transsexual', *Archives of General Psychiatry*, 28 (1973), 569–76, quote at 569.

150 D. H. Barlow, G. G. Abel, and E. B. Blanchard, 'Gender Identity Change in Transsexuals: Follow-Up and Replications', *Archives of General Psychiatry*, 36 (1979), 1001–7, quote at 1001.

151 S. Fredericks, 'Aversion Therapy: TV Cure or Psychic Vivisection?' *Turnabout: A Magazine of Transvestism*, 1 (1963), 3–6, quote at 4.

152 Barlow, Abel, and Blanchard, 'Gender Identity Change in Transsexuals: Follow-Up and Replications', 1001, 1002.

153 Barlow, Reynolds, and Agras, 'Gender Identity Change in a Transsexual', 569, 570.

154 Ibid., 571.

155 Ibid.

156 Ibid.

157 Ibid.

158 Ibid., 572.

159 Ibid.

160 Ibid.

161 Ibid., 573.

162 Ibid.

163 Ibid.

164 Ibid.

165 Ibid., 574.

166 Barlow, Abel, and Blanchard, 'Gender Identity Change in Transsexuals: Follow-Up and Replications', 1004.

167 Ibid.

168 Ibid., 1001.

169 Ibid., 1002.

170 Ibid., 1005, 1006.

171 H. J. Baker, and R. Green, 'Treatment of Transsexualism', *Current Psychiatric Therapies*, 10 (1970), 88–99, quote at 91.

172 Barlow, Abel, and Blanchard, 'Gender Identity Change in Transsexuals: Follow-Up and Replications', 1001.

173 Dickinson, *Curing Queers*, pp. 76, 181–3, 184–5, 186–8.

174 J. N. Marquis, 'Orgasmic Reconditioning: Changing Sexual Object Choice Through Controlling Masturbation Fantasies', *Journal of Behavioral Therapy and Experimental Psychiatry*, 1:4 (1970), 263–71.

175 J. G. Thorpe and E. Schmidt, 'Therapeutic Failure in a Case of Aversion Therapy', *Behaviour Research and Therapy*, 1:2–4 (1963), 293–6, quote at 294.

176 Lavin and others, 'Behavior Therapy in a Case of Transvestism', 352; J. C. Barker, 'Behaviour Therapy for Transvestism: A Comparison of Pharmacological and Electrical Aversion Techniques', *British Journal of Psychiatry*, 111 (1965), 268–76, at 271.

177 Blakemore and others, 'Application of Faradic Aversion Conditioning in a Case of Transvestism', 33–4.

178 M. P. Feldman, 'Aversion Therapy for Sexual Deviations: A Critical Review', *Psychological Bulletin*, 65:2 (1966), 65–79, at 71.

179 D. H. Barlow and W. S. Agras, 'Fading to Increase Heterosexual Responsiveness in Homosexuals', *Journal of Applied Behavior Analysis*, 6:3 (1973), 355–66, at 358.

180 H. Brierley, *Transvestism: A Handbook with Case Studies for Psychologists, Psychiatrists and Counsellors* (Oxford, 1979), p. 195.

181 L. M. Lothstein, 'Psychotherapy with Patients with Gender Dysphoria Syndromes', *Bulletin of the Menninger Clinic*, 41:6 (1977), 563–82, quote at 564.

182 Lothstein, *Female-to-Male Transsexualism*, p. 65.

183 Ibid., p. 76.

184 J. K. Meyer, 'Clinical Variants Among Applicants for Sex Reassignment', *Archives of Sexual Behavior*, 3:6 (1974), 527–58, quote at 527.

185 Ibid., 553.

186 Ibid., 534.

187 Ibid., 544.

188 Ibid., 544, 546.

189 Ibid., 554–5.

190 Ibid., 556.

191 Not always with the approval of Johns Hopkins: ibid., 527–58.

192 D. R. Laub and N. Fisk, 'A Rehabilitation Program for Gender Dysphoria Syndrome by Surgical Sex Change', *Plastic and Reconstructive Surgery*, 53:4 (1974), 388–403, esp. 389, 398, 401.

193 Ibid., 388, 401.

194 N. Fisk, 'Gender Dysphoria Syndrome (The How, What, and Why of a Disease)', in Laub and Gandy (eds.), *Proceedings*, pp. 7–14, quotes at p. 8.

195 Ibid., p. 9.

196 Ibid., p. 10.

197 Kubie and Mackie, 'Critical Issues', 436.

198 D. W. Hastings, 'Postsurgical Adjustment of Male Transsexual Patients', *Clinics in Plastic Surgery*, 1:2 (1974), 335–44, quote at 344.

199 Ibid., 340–1.

200 Ibid., 344.

201 T. Kando, *Sex Change: The Achievement of Gender Identity Among Feminized Transsexuals* (Springfield, Ill., 1973), pp. 9–12.

202 Ibid., pp. 11–12.

203 Ibid., pp. 9, 12–14 (quote at 14).

204 Ibid., p. 10.

205 Ibid., pp. 49, 60 (quote at 60).

206 W. B. Pomeroy, 'Transsexualism and Sexuality: Sexual Behavior of Pre- and Postoperative Male Transsexuals', in Green and Money (eds.), *Transsexualism and Sex Reassignment*, ch. 10.

207 Ibid., p. 188.

208 Ibid.

209 Ibid.

210 R. Green, 'Psychiatric Management of Special Problems in Transsexualism', in Green and Money (eds.), *Transsexualism and Sex Reassignment*, ch. 19, quote at p. 288.

211 Ibid., pp. 283–4.

212 Ibid.

213 J. K. Meyer and J. E. Hoopes, 'The Gender Dysphoria Syndromes: A

Position on So-Called "Transsexualism"', *Plastic and Reconstructive Surgery*, 54:4 (1974), 444–51, quote at 450.

214 Laub and Fisk, 'A Rehabilitation Program', 388–403, quotes at 400.

215 Ibid., 400.

216 Ibid., 388.

217 Stoller, *Transsexual Experiment*, p. 81.

218 Ibid., p. 89.

219 Ibid., p. 90.

220 Ibid., p. 83.

221 Lothstein, *Female-to-Male Transsexualism*, pp. 72–3.

222 K. R. MacKenzie, 'Gender Dysphoria Syndrome: Towards Standardized Diagnostic Criteria', *Archives of Sexual Behavior*, 7:4 (1978), 251–62, quotes at 251, 252.

223 V. Prince, 'Transsexuals and Pseudotranssexuals', *Archives of Sexual Behavior*, 7:4 (1978), 263–72, quotes at 271.

224 A. J. Morgan, 'Psychotherapy for Transsexual Candidates Screened Out of Surgery', *Archives of Sexual Behavior*, 7:4 (1978), 273–83, quotes at 274, 283.

225 Ibid., 279.

226 J. M. Noe and othes, 'Construction of Male Genitalia: The Stanford Experience', *Archives of Sexual Behavior*, 7:4 (1978), 297–303.

227 B. N. Jayaram, O. H. Stuteville, and I. M. Bush, 'Complications and Undesirable Results of Sex-Reassignment Surgery in Male-to-Female Transsexuals', *Archives of Sexual Behavior*, 7:4 (1978), 337–45, quote at 340.

228 See Figures 1–12 in ibid., 342–5.

229 For mobility, see Jayaram, Stuteville, and Bush, 'Complications and Undesirable Results', 338; D. Hastings and C. Markland, 'Post-Surgical Adjustment of Twenty-Five Transsexuals (Male-to-Female) in the University of Minnesota Study', *Archives of Sexual Behavior*, 7:4 (1978), 327–36, at 327; 'Discussion', *Archives of Sexual Behavior*, 7:4 (1978), 385–6.

230 'Open Forum', *Archives of Sexual Behavior*, 7:4 (1978), 387–415, quote at 397.

231 Ibid., 393.

232 N. M. Fisk, 'Five Spectacular Results', *Archives of Sexual Behavior*, 7:4 (1978), 351–69, quote at 353.

233 Morgan, 'Psychotherapy for Transsexual Candidates', 276.

234 'Discussion', *Archives of Sexual Behavior*, 7:4 (1978), 313–14.

235 Meyerowitz, *How Sex Changed*, pp. 266–70.

236 J. E. Brody, 'Benefits of Transsexual Surgery Disputed as Leading Hospital Halts the Procedure', *New York Times*, 2 October 1979.

237 P. R. McHugh, 'Surgical Sex: Why We Stopped Doing Sex Change Operations', *First Things*, November 2004: www.firstthings.com/article/2004/11/surgical-sex.

238 P. R. McHugh, 'Psychiatric Misadventures', *The American Scholar*, 61:4 (1992), 497–510, quote at 498.

239 S. J. Kessler and W. McKenna, *Gender: An Ethnomethodological Approach* (New York, 1978), p. 118.

240 D. B. Billings and T. Urban, 'The Socio-Medical Construction of Transsexualism: An Interpretation and Critique', *Social Problems*, 29:3 (1982), 266–82, quote at 266.

241 Ibid., 275–6.

242 Ibid., 275.

243 Ibid., 273–4.

244 Ibid., 276.

245 Meyer, 'Theory of Gender Identity Disorders', 384.

246 M. E. Petersen and R. Dickey, 'Surgical Sex Reassignment: A Comparative Survey of International Centers', *Archives of Sexual Behavior*, 24:2 (1995), 135–56 (149 for the 1981 allegation).

247 C. Millot, *Horsexe: Essays on Transsexuality*, trans. K. Hylton (New York, 1990), pp. 107–8. First published in French in 1983.

248 Richards, *Second Serve*, p. 163.

249 GLBT Historical Society, San Francisco, Periodical Collection: *Drag: A Magazine About the Transvestite*, 7:25 (1977).

250 Ibid.

5 The Transgender Turn

1 N. Krieger, 'Writing Trans', in L. Erickson-Schroth (ed.), *Trans Bodies, Trans Selves: A Resource for the Transgender Community* (New York, 2014), pp. 582–3, quote at p. 583.

2 Ibid, p. 583.

3 N. Krieger, *Nina Here Nor There: My Journey Beyond Gender* (Boston, Mass., 2011), p. 91.

4 Krieger, 'Writing Trans', p. 583.

5 B. J. Noble, 'Our Bodies Are Not Ourselves: Tranny Guys and the Racialized Class Politics of Incoherence', in S. Stryker and A. Z. Aizura (eds.), *The Transgender Studies Reader 2* (New York, 2013), ch. 20, quotes at p. 253.

6 S. Stryker, 'Perfect Day', in T. O'Keefe and K. Fox (eds.), *Trans People in Love* (New York, 2008), ch. 5, quote at p. 50.

7 K. Bornstein, *Gender Outlaw: On Men, Women, and the Rest of Us* (New York, 1995), pp. 3–4. First published in 1994.

8 J. Baudrillard, 'Transsexuality', in his *The Transparency of Evil: Essays on Extreme Phenomena*, trans. J. Benedict (London, 1993), p. 21.

9 R. F. Docter, *Transvestites and Transsexuals: Toward a Theory of Cross-Gender Behavior* (New York, 1990), p. 21. First published in 1988.

10 V. Prince, 'Change of Sex or Gender', *Transvestia*, 60 (1969), 53–65, quote at 65; V. Prince, 'These Eventful Years', *Transvestia*, 61 (1970), 3–22, quote at 20.

11 Docter, *Transvestites and Transsexuals*, pp. 21–2.

12 M. Mieli, *Towards a Gay Communism: Elements of a Homosexual Critique* (London, 2018), pp. 6, 8. First published as *Homosexuality and Liberation* in 1980.

13 C. Williams, 'Transgender', *TSQ*, 1:1–2 (2014), 232–4.

14 See J. Meyerowitz, *How Sex Changed: A History of Transsexuality in the United States* (Cambridge, Mass., 2002), ch. 7. For the apt phrase about 'heterosexual graduates', see D. Denny, 'Changing Models of Transsexualism', *Journal of Gay & Lesbian Psychotherapy*, 8:1–2 (2004), 25 40, quote at 29.

15 Docter, *Transvestites and Transsexuals*, p. 37.

16 S. Stryker and P. Currah, 'Introduction', *TSQ*, 1:1–2 (2014), 1–18, quote at 5.

17 D. Denny, 'Transgender Communities of the United States in the Late Twentieth Century', in P. Currah, R. M. Juang, and S. P. Minter (eds.), *Transgender Rights* (Minneapolis, 2006), ch. 9, quote at p. 184.

18 H. Boswell, 'The Transgender Alternative', *Chrysalis Quarterly*, 1:2 (1991), 29–31.

19 L. Feinberg, *Transgender Liberation: A Movement Whose Time Has Come* (New York, 1992), p. 5.

20 Bornstein, *Gender Outlaw*, pp. 12–13.

21 G. O. MacKenzie, *Transgender Nation* (Bowling Green, OH, 1994), p. 11.

22 Z. I. Nataf, *Lesbians Talk Transgender* (London, 1996).

23 H. Boswell, 'The Transgender Paradigm Shift Toward Free Expression', in B. Bullough, V. L. Bullough, and J. Elias (eds.), *Gender Blending* (Amherst, NY, 1997), pp. 53–7, quote at p. 56.

24 M. P. Allen, *The Gender Frontier* (Heidelberg, 2003); D. L. Volcano, *Sublime Mutations* (Tübingen, 2000).

25 J. Butler, *Gender Trouble: Feminism and the Subversion of Identity* (New York, 1990), pp. 6–7, 17, 25, 139.

26 Ibid, p. 112.

27 Ibid, p. 127.

28 For just some examples, see J. L. Reich, 'Genderfuck: The Law of the Dildo', *Discourse*, 15:1 (1992), 112–27; R. Felski, 'Fin de Siècle, Fin de Sexe: Transsexuality, Postmodernism, and the Death of History', *New Literary History*, 27:2 (1996), 337–49; C. Straayer, *Deviant Eyes, Deviant Bodies: Sexual Re-Orientations in Film and Video* (New York, 1996), pp. 253–87; C. J. Hale, 'Leatherdyke Boys and Their Daddies: How to Have Sex Without Women or Men', *Social Text*, 52/3 (1997), 223–36; K. Shaffer, 'The Game Girls of VNS Matrix: Challenging Gendered Identities in Cyberspace', in M. A. O'Farrell and L. Vallone (eds.), *Virtual Gender: Fantasies of Subjectivity and Embodiment* (Ann Arbor, Mich., 1999), pp. 147–68.

29 Ekins and King (eds.), *Blending Genders*, chs. 3, 14. See also L. Cameron, *Body Alchemy: Transsexual Portraits* (San Francisco, 1996).

30 As does Denny, 'Transgender Communities', p. 184, and in an earlier intervention: D. Denny, 'The Politics of Diagnosis and a Diagnosis of Politics: The University-Affiliated Gender Clinics, and How They Failed to Meet the Needs of Transsexual People', *Chrysalis Quarterly*, 1:3 (1992), 9–20.

31 D. Levy, 'Two Transsexuals Reflect on University's Pioneering Gender Program', *Stanford Report*, 3 May 2000: http://news.stanford.edu/news/2000/may3/sexchange-53.html.

32 M. E. Petersen and R. Dickey, 'Surgical Sex Reassignment: A Comparative Survey of International Centers', *Archives of Sexual Behavior*, 24:2 (1995), 135–56, esp. 151–2.

33 *Diagnostic and Statistical Manual of Mental Disorders (Third Edition): DSM-III* (Washington, DC, 1980), pp. 261–6; *Diagnostic and Statistical Manual of Mental Disorders (Fourth Edition): DSM-IV* (Washington, DC, 1994), pp. 493, 532–8; *Diagnostic and Statistical Manual of Mental Disorders (Fourth Edition, Text Revision): DSM-IV-TR* (Arlington, Va., 2000), pp. 535, 576–82; *Diagnostic and Statistical Manual of Mental Disorders (Fifth Edition): DSM-5* (Washington, DC, 2013), pp. 451–9. DSM-III had a category Gender Identity Disorder of Childhood (GIDC) (pp. 264–6), which was incorporated into Gender Identity Disorder in DSM-IV and IV-TR, but then separated again in DSM-5 into Gender Dysphoria in Children and Gender Dysphoria in Adolescents and Adults (pp. 453–3). DSM-III-R (1987) represented a transitional stage in this shift in nomenclature, with its four Gender Identity Disorders: Gender Identity Disorder of Childhood; Transsexualism; Gender Identity Disorder of Adolescence or Adulthood, Nontranssexual Type; and Gender Identity Disorder Not Otherwise Specified. See *Diagnostic and Statistical Manual of Mental Disorders (Third Edition - Revised): DSM-III-R* (Washington, DC, 1987), pp. 71–8. For a perceptive account of the history of GIDC before DSM-5, see K. Bryant, 'Making Gender Identity Disorder of Childhood: Historical Lessons for Contemporary Debates', *Sexuality Research & Social Policy*, 3:3 (2006), 23–39; and K. Bryant, 'Diagnosis and Medicalization', *Advances in Medical Sociology*, 12 (2011), 33 57.

34 J. W. Barnhill, 'Gender Dysphoria: Introduction', in J. W. Barnhill (ed.), *DSM-5 Clinical Cases* (Washington, DC, 2014), ch. 14, quote at p. 238.

35 B. W. Steiner (ed.) *Gender Dysphoria: Development, Research, Management* (New York, 1985).

36 J. M. Cantor and K. S. Sutton, 'Paraphilia, Gender Dysphoria, and Hypersexuality', in P. H. Blaney, R. F. Krueger, and T. Millon

(eds.), *Oxford Textbook of Psychopathology, Third Edition* (New York, 2015), ch. 22.

37 A. H. Johnson, 'Normative Accountability: How the Medical Model Influences Transgender Identities and Experiences', *Sociology Compass*, 9:9 (2015), 803–13, quotes at 804.

38 Z. Davy, 'The DSM-5 and the Politics of Diagnosing Transpeople', *Archives of Sexual Behavior*, 44:5 (2015), 1165–76, at 1165. For a methodical illustration of this, see Z. Davy and M. Toze, 'What is Gender Dysphoria? A Critical Systematic Narrative Review', *Transgender Health*, 3:1 (2018), 159–69.

39 A. I. Lev, 'Gender Dysphoria: Two Steps Forward, One Step Back', *Clinical Social Work Journal*, 41:3 (2013), 288–96.

40 J. Drescher, 'Transsexualism, Gender Identity Disorder and the DSM', *Journal of Gay & Lesbian Health*, 14:2 (2010), 109–22; *DSM-5*, p. 451; Barnhill (ed.), *DSM-5 Clinical Cases*, p. 238.

41 D. King, 'Gender Blending: Medical Perspectives and Technology', in R. Ekins and D. King (eds.), *Blending Genders: Social Aspects of Cross-Dressing and Sex-Changing* (London, 1996), ch. 7, esp. pp. 95–8.

42 Denny, 'Transgender Communities', p. 184.

43 C. Griggs, *S/he: Changing Sex and Changing Clothes* (New York, 2003), pp. 1, 92. Emphasis in original. First published in 1998.

44 L. B. Girshick, *Transgendered Voices: Beyond Woman and Men* (Lebanon, NH, 2008), p. 1.

45 M. Nelson, *The Argonauts* (Melbourne, 2016), Ebook, loc. 751. First published in 2015.

46 P. Preciado, *Testo Junkie: Sex, Drugs, and Biopolitics in the Pharmacopornographic Era* (New York, 2013), p. 16.

47 Ibid., p. 263.

48 C. BrianKate, 'Paradox is Paradise for Me', in T. O'Keefe and K. Fox (eds.), *Finding the Real Me: True Tales of Sex and Gender Diversity* (San Francisco, 2003), pp. 1–10, quote at p. 1.

49 P. A. Bernhardt-House, 'So, Which One is the Opposite Sex?', in O'Keefe and Fox (eds.), *Finding the Real Me*, pp. 76–87.

50 P. B. Preciado, *Counter-Sexual Manifesto*, trans. K. G. Dunn (New York, 2018), pp. 17, 21.

51 W. O. Bockting, 'Psychotherapy and the Real-Life Experience: From Gender Dichotomy to Gender Diversity', *Sexologies*, 17:4 (2008), 211–24.

52 Z. Davy, *Recognizing Transsexuals: Personal, Political and Medicolegal Embodiment* (Farnham, 2011), ch. 3, quote at p. 95.

53 M. Heinz, *Entering Masculinity: The Inevitability of Discourse* (Chicago, 2016), p. 3.

54 T. P. McBee, *Amateur: A True Story About What Makes a Man* (New York, 2018), p. 3.

55 Ibid., p. 38.

56 G. Hansbury, 'King Kong & Goldilocks: Imagining Transmasculinities Through the Trans–Trans Dyad', *Psychoanalytic Dialogues*, 21:2 (2011), 210–20, quote at 213.

57 C. J. Williams, M. S. Weinberg, and J. G. Rosenberger, 'Trans Men: Embodiments, Identities, and Sexualities', *Sociological Forum*, 28:4 (2013), 719–41.

58 C. J. Williams, M. S. Weinberg, and J. G. Rosenberger, 'Trans Women Doing Sex in San Francisco', *Archives of Sexual Behavior*, 45:7 (2016), 1665–78.

59 A. Sprinkle, 'My First Female-to-Male Transsexual Lover', in M. Christian (ed.), *Transgender Erotica* (New York, 2006), pp. 17–25, quote at p. 23.

60 K. Schilt and E. Windsor, 'The Sexual Habitus of Transgender Men: Negotiating Sexuality Through Gender', *Journal of Homosexuality*, 61:5 (2014), 732–48.

61 S. Lowrey, 'Made Real', in M. Diamond (ed.), *Trans/Love: Radical Sex, Love, and Relationships Beyond the Gender Binary* (San Francisco, 2011), Ebook, loc. 1575.

62 H. Devor, *FTM: Female-to-Male Transsexuals in Society* (Bloomington and Indianapolis, 1999), p. 483.

63 H. Devor, 'Sexual Orientation Identities, Attractions, and Practices of Female-to-Male Transsexuals', *The Journal of Sex Research*, 30:4 (1993), 303–15, quote at 304.

64 Devor, *FTM*, pp. 501–10.

65 Ibid., p. 311.

66 See S. Stryker, 'Portrait of a Transfag Drag Hag as a Young Man: The Activist Career of Louis G. Sullivan', in K. More and

S. Whittle (eds.), *Reclaiming Genders: Transsexual Grammars at the Fin de Siècle* (New York, 1999), ch. 3; L. M. Rodemeyer, *Lou Sullivan Diaries (1970–1980) and Theories of Sexual Embodiment: Making Sense of Sensing* (New York, 2017); B. Reay, 'A Transgender Story: The Diaries of Louis Graydon Sullivan', in B. Reay, *Sex in the Archives: Writing American Sexual Histories* (Manchester, 2018), ch. 5.

67 Petersen and Dickey, 'Surgical Sex Reassignment', 135–56, esp. 145.

68 D. Schleifer, 'Make Me Feel Mighty Real: Gay Female-to-Male Transgenderists Negotiating Sex, Gender, and Sexuality', *Sexualities*, 9:1 (2006), 57–75, quote at 57–8.

69 Ibid., 71.

70 Ibid., 72.

71 W. Bockting, A. Benner, and E. Coleman, 'Gay and Bisexual Identity Development Among Female-to-Male Transsexuals in North America: Emergence of a Transgender Sexuality', *Archives of Sexual Behavior*, 38:5 (2009), 688–701, quote at 693.

72 Ibid., 694–5.

73 Ibid., 696.

74 Ibid., 697.

75 P. Gagné and R. Tewksbury, 'Knowledge and Power, Body and Self: An Analysis of Knowledge Systems and the Transgendered Self', *The Sociological Quarterly*, 40:1 (1999), 59–83, quote at 79.

76 N. C. Barnes, 'Kate Bornstein's Gender and Genre Bending', in L. Senelick (ed.), *Gender in Performance: The Presentation of Difference in the Performing Arts* (Hanover, NH, 1992), pp. 311–23, quote at p. 322.

77 J. Cromwell, *Transmen & FTMs: Identities, Bodies, Genders & Sexualities* (Urbana and Chicago, 1999), p. 25.

78 Ibid., p. 28.

79 Ibid., pp. 28–30.

80 Ibid., p. 131.

81 M. Diamond (ed.), *From the Inside Out: Radical Gender Transformation, FTM and Beyond* (San Francisco, 2004).

82 M. Diamond, 'Breaking the Gender Mold', in ibid., p. 8.

83 B. Potential, 'Monster Trans', in Diamond (ed.), *From the Inside Out*, p. 33.

84 M. R. Van, 'Thoughts on Transcending Stone: The Tale of One Transgendered Man and His Journey to Find Sexuality in His New Skin', in Diamond (ed.), *From the Inside Out*, p. 55.

85 R. Vanderburgh, 'Living la Vida Medea', in Diamond (ed.), *From the Inside Out*, p. 106.

86 J.-A. Lamas, 'GenderFusion', in Diamond (ed.), *From the Inside Out*, p. 120.

87 G. A. L. Ansara, 'Transitioning *or* What's a Nice Dyke Like Me Doing Becoming a Gay Man?', in Diamond (ed.), *From the Inside Out*, p. 91.

88 G. Beemyn and S. Rankin, *The Lives of Transgender People* (New York, 2011); J. Green, D. Denny, and J. Cromwell, '"What Do You Want Us to Call You?" Respectful Language', *TSQ*, 5:1 (2018), 100–10.

89 Beemyn and Rankin, *Lives of Transgender People*, pp. 23–6.

90 Ibid., p. 27.

91 Ibid., pp. 32–3.

92 L. E. Kuper, R. Nussbaum, and B. Mustanski, 'Exploring the Diversity of Gender and Sexual Orientation Identities in an Online Sample of Transgender Individuals', *Journal of Sex Research*, 49:2–3 (2012), 244–54, quote at 246.

93 Ibid., 247.

94 Ibid.

95 Ibid., 251.

96 J. Sevelius, '"There's No Pamphlet for the Kind of Sex I Have": HIV-Related Risk Factors and Protective Behaviors Among Transgender Men Who Have Sex with Nontransgender Men', *Journal of the Association of Nurses in AIDS Care*, 20:5 (2009), 398–410.

97 Ibid., 404.

98 Ibid.

99 Ibid., 402.

100 See, for example, M. Warner (ed.), *Fear of a Queer Planet: Queer Politics and Social Theory* (Minneapolis, 1993); A. Jagose, *Queer Theory* (Dunedin, 1996).

101 L. Zimman, 'Transmasculinity and the Voice: Gender Assignment, Identity, and Presentation', in T. M. Milani (ed.), *Language and Masculinities: Performances, Intersections, Dislocations* (New York, 2015), ch. 10, quotes from p. 202.

102 D. Troka, K. Lebesco, and J. Noble (eds.), *The Drag King Anthology* (New York, 2002).

103 J. J. Halberstam, 'What is a Drag King?', in D. L. Volcano and J. J. Halberstam (eds.), *The Drag King Book* (London, 1999), ch. 1, quotes at p. 36.

104 See the essays by S. P. Schacht and J. L. Patterson in Troka, Lebesco, and Noble (eds.), *The Drag King Anthology*.

105 Halberstam, 'What is a Drag King?', p. 36.

106 D. L. Volcano, '"A Kingdom Comes"', in Volcano and Halberstam (eds.), *The Drag King Book*, Foreword, quote at p. 27.

107 L. J. Rupp, V. Taylor, and E. I. Shapiro, 'Drag Queens and Drag Kings: The Difference Gender Makes', *Sexualities*, 13:3 (2010), 275–94, quotes at 282.

108 K. Rosenfeld, 'Drag King Magic: Performing the Other', in Troka, Lebesco, and Noble (eds.), *The Drag King Anthology*, pp. 201–19, quote at p. 217.

109 Volcano and Halberstam (eds.), *The Drag King Book*; G. Baur, *Venus Boyz* (2001).

110 Fales Library Special Collections, Elmer Holmes Bobst Library, New York University, MSS 268, Johnny Science Papers, 1948–2000 (hereafter Johnny Science Papers).

111 Ibid., Series 1: Projects, Box 2, Folder 3, Flyer.

112 K. R. Horowitz, 'The Trouble with "Queerness": Drag and the Making of Two Cultures', *Signs*, 38:2 (2013), 303–26, quote at 311.

113 Halberstam, 'What is a Drag King?', p. 39.

114 J. Halberstam, 'Interview with a Drag King #2: The Dodge Bros, San Francisco, 1997', in Volcano and Halberstam (eds.), *The Drag King Book*, ch. 6, quote at p. 132.

115 See Kirsty MacDonald's documentary short film *Blending the Female and Male Through MilDred* (2010): www.youtube.com/watch?v=pWAg3DsEnaA. There is also an extended discussion of Gerestant in M. E. Alías, *Long Live the King: A Genealogy of Performative Genders* (Newcastle upon Tyne, 2009), pp. 134–67.

116 P. Shaw, 'On Being an Independent Solo Artist (No Such Thing)', in J. Dolan (ed.), *A Menopausal Gentleman: The Solo Performances of Peggy Shaw* (Ann Arbor, Mich., 2011), pp. 39–41, quote at p. 41.

117 Scattered throughout Volcano and Halberstam (eds.), *The Drag King Book*.

118 D. L. Volcano and J. Halberstam, 'Interview with a Drag King #1: Mo B. Dick, New York City, 1996', in ibid., ch. 4, quote at p. 114. For an example of his performance, see https://www.youtube.com/watch?v=6AgvuJqQXnI.

119 Halberstam, 'Class, Race and Masculinity: The Superfly, the MacDaddy and the Rapper', in Volcano and Halberstam (eds.), *The Drag King Book*, ch. 7.

120 See M. E. Alías, 'Shattering Gender Taboos in Gabriel Bauer's *Venus Boyz*', *Journal of Gender Studies*, 19:2 (2010), 167–79.

121 E. Shapiro, 'Drag Kinging and the Transformation of Gender Identities', *Gender & Society*, 21:2 (2007), 250–71, quote at 259.

122 K. Davy, *Lady Dicks and Lesbian Brothers: Staging the Unimaginable at the WOW Café Theatre* (Ann Arbor, Mich., 2011), pp. 8, 88, 159, and illustrations between pages 178 and 179.

123 D. Torr and S. Bottoms, *Sex and Drag and Male Roles: Investigating Gender as Performance* (Ann Arbor, Mich., 2010), p. 26.

124 Johnny Science Papers, Series 1: Projects, Box 2, Folder 15.

125 Ibid., Series 3: Photography, Box 5, Folder 40: Drag King Club photographs.

126 Ibid., Series 3: Photography, Binder 1: File: FTM Fraternity: Drag King Club Photography 2 – *c*.1995.

127 Ibid., Series 1: Projects, Box 1, Folder 36: Drag Kings: Flyers, *c*.1996–1998; Box 2, Folder 1: F2M Fraternity Administrative File.

128 N. Goldin, *The Other Side* (Zurich, 1995), p. 7.

129 M. Economy and J. Wandrag, *Pansy Beat* (New York, 2017); J. Fleisher, *The Drag Queens of New York: An Illustrated Field Guide* (London, 1997).

130 See A. Badertscher, *Baltimore Portraits* (Durham, NC, 1999). The Badertscher archive is divided between his residence in Baltimore (hereafter BB) and the Leslie–Lohman Museum of Gay and Lesbian Art in New York. The figures reproduced here are from

high-quality digital images of the original photographs from Badertscher's personal archive.

131 A. Badertscher, 'Biographies of Models', BB.

132 John Flowers Portfolio, BB.

133 V. Taylor and L. J. Rupp, 'Chicks with Dicks, Men in Dresses: What It Means to Be a Drag Queen', *Journal of Homosexuality*, 46:3/4 (2004), 113–33, quotes at 121, 122.

134 Ibid., 122.

135 L. J. Rupp and V. Taylor, *Drag Queens at the 801 Cabaret* (Chicago, 2003), p. 32.

136 S. Fenstermaker and N. Jones, 'V. Taylor and L. J. Rupp', in S. Fenstermaker and N. Jones, *Sociologists Backstage: Answers to 10 Questions About What They Do* (New York, 2011), ch. 17, quote at pp. 218–19.

137 R. Barrett, *From Drag Queens to Leathermen: Language, Gender, and Gay Male Subcultures* (New York, 2017), p. 36.

138 Ibid.

139 Rupp and Taylor, *Drag Queens*, p. 126.

140 Ibid., p. 134.

141 Ibid., p. 137.

142 Ibid., p. 201.

143 J. Egner and P. Maloney, '"It Has No Color, It Has No Gender, It's Gender Bending": Gender and Sexual Fluidity and Subversiveness in Drag Performance', *Journal of Homosexuality*, 63:7 (2016), 875–903, quote at 898.

144 C.-A. Taylor, 'Boys Will be Girls: The Politics of Gay Drag', in D. Fuss (ed.), *Inside/Out: Lesbian Theories, Gay Theories* (New York, 1991), ch. 2, quote at p. 42.

145 Butler, *Gender Trouble*, p. 137. The original words were in italics but I have dispensed with those.

146 J. Butler, *Undoing Gender* (New York, 2004), p. 217.

147 J. E. Muñoz, *Disidentifications: Queers of Color and the Performance of Politics* (Minneapolis, 1999), ch. 4, quotes at pp. 100, 115.

148 See A. Z. Aizura and others (eds.), 'Decolonizing the Transgender Imaginary', *TSQ*, 1:3 (2014).

149 See discussions in W. Roscoe, *The Zuni Man–Woman* (Albuquerque, NM, 1991), pp. 211–13; S. Lang, 'Various Kinds of Two-Spirit

People: Gender Variance and Homosexuality in Native American Communities', in S.-E. Jacobs, W. Thomas, and S. Lang (eds.), *Two-Spirit People: Native American Gender Identity, Sexuality, and Spirituality* (Urbana, Ill., 1997), ch. 4; J. Roscoe, *Changing Ones: Third and Fourth Genders in Native America* (New York, 1998), esp. chs. 1, 6.

150 W. L. Williams, *The Spirit and the Flesh: Sexual Diversity in American Indian Culture* (Boston, Mass., 1992), p. 85.

151 For the quote, see Roscoe, *Changing Ones*, p. 4.

152 D. Kulick, *Travesti: Sex, Gender and Culture Among Brazilian Transgendered Prostitutes* (Chicago, 1998), p. 6 (for quote); M. R. V. Garcia, 'Identity as a "Patchwork": Aspects of Identity Among Low-Income Brazilian *Travestis*', *Culture, Health & Sexuality*, 11:6 (2009), 611–23.

153 A. Prieur, *Mema's House, Mexico City: On Transvestites, Queens, and Machos* (Chicago, 1998), p. 267.

154 A. Dutta and R. Roy, 'Decolonizing Transgender in India: Some Reflections', *TSQ*, 1:3 (2014), 320–37, quotes at 321, 330. See, also, G. Reddy, *With Respect to Sex: Negotiating Hijra Identity in South India* (Chicago, 2005).

155 S. Hines, 'Sexing Gender; Gendering Sex: Towards an Intersectional Analysis of Transgender', in Y. Taylor, S. Hines, and M. E. Casey (eds.), *Theorizing Intersectionality and Sexuality* (New York, 2011), ch. 7; J. L. Nagoshi, C. T. Nagoshi, and S. Brzuzy, 'Intersectionality and Narratives of Lived Experience', in their *Gender and Sexual Identity: Transcending Feminist and Queer Theory* (New York, 2014), ch. 7.

156 E. Skidmore, 'Constructing the "Good Transsexual": Christine Jorgensen, Whiteness, and Heteronormativity in the Mid-Twentieth-Century Press', *Feminist Studies*, 37:2 (2011), 270–300.

157 Sylvia Rivera in S. Rivera and M. P. Johnson, *Street Transvestite Action Revolutionaries: Survival, Revolt, and Queer Antagonist Struggle* (Untorelli Press [no place of publication], 2013), p. 30.

158 The cat analogy is Marsha Johnson's, in Rivera and Johnson, *Street Transvestite Action Revolution*, p. 26.

159 Rivera and Johnson, *Street Transvestite Action Revolution*, p. 35.

160 T. Raun, *Out Online: Trans Self-Representation and Community Building on YouTube* (London, 2016), pp. 8, 90–5.

161 Heinz, *Entering Masculinity*, p. 86.

162 *Still Black: A Portrait of Black Transmen* (2008: dir. K. R. Ziegler); J. Tyburczy, 'Interview with K. R. Ziegler', *Trans Media: Still Black*, University of California TV, 31 June 2017: www.uctv.tv/shows/ Still-Black-A-Portrait-of-Black-Transmen-Trans-Media-32541.

163 For Cowen, see The New-York Historical Society, Department of Prints, Photographs, and Architectural Collections, PR 050 Photographer File, Box 5, Folder: Jeff Cowen; J. Festa, 'The Drag Queen Stroll: Jeff Cowen and 1980s New York City', 11 September 2013: http://blog.nyhistory.org/the-drag-queen-stroll-jeff-cowen-and-1980s-new-york-city. For Naito, see K. Naito, *West Side Rendezvous* (London, 2011).

164 Naito, *West Side Rendezvous*.

165 *The Salt Mines* (1990: dir. S. Aikin and C. Aparicio).

166 *The Aggressives* (2005: dir. D. Peddle).

167 J. Halberstam, 'Global Female Masculinities', *Sexualities*, 15:3–4 (2012), 336–54, quote at 349.

168 J. W. Wright, *Trans/Portraits: Voices from Transgender Communities* (Hanover, NH, 2015), pp. 104–5, 113.

169 S. J. Hwahng and L. Nuttbrock, 'Sex Workers, Fem Queens, and Cross Dressers: Differential Marginalizations and HIV Vulnerabilities Among Three Ethnocultural Male-to-Female Transgender Communities in New York City', *Sexuality Research & Social Policy*, 4:4 (2007), 36–59.

170 K. M. Brown, 'Mimesis in the Face of Fear: Femme Queens, Butch Queens, and Gender Play in the Houses of Greater Newark', in M. C. Sánchez and L. Schlossberg (eds.), *Passing: Identity and Interpretation in Sexuality, Race, and Religion* (New York, 2001), ch. 8.

171 D. Rowan, D. D. Long, and D. Johnson, 'Identity and Self-Presentation in the House/Ball Culture: A Primer for Social Workers', *Journal of Gay & Lesbian Social Services*, 25:2 (2013), 178–96, quote at 185.

172 Ibid. See, also, M. M. Bailey, 'Performance as Intravention: Ballroom Culture and the Politics of HIV/AIDS in Detroit', *Souls*, 11:3 (2009), 253–74; M. M. Bailey, 'Gender/Racial Realness:

Theorizing the Gender System in Ballroom Culture', *Feminist Studies*, 37:2 (2011), 365–86; and M. M. Bailey, *Butch Queens Up in Pumps: Gender, Performance, and Ballroom Culture in Detroit* (Ann Arbor, Mich., 2013).

173 Bailey, 'Gender/Racial Realness', 370–1.

174 Ibid., 371.

175 Bailey, *Butch Queens Up in Pumps*, p. 37.

176 L. Hilderbrand, *Paris Is Burning: A Queer Film Classic* (Vancouver, 2013), p. 23.

177 G. H. Gaskin, *Legendary: Inside the House Ballroom Scene* (Durham, NC, 2013), no pagination.

178 Rivera and Johnson, *Street Transvestite Action Revolution*, p. 48. Emphasis in original.

179 Bailey, 'Gender/Racial Realness', 375.

180 Hilderbrand, *Paris Is Burning*, p. 72.

181 *Paris Is Burning* (1990/1: dir. J. Livingston); Hilderbrand, *Paris Is Burning*, p. 76.

182 *Paris Is Burning* (1990/1); Hilderbrand, *Paris Is Burning*, p. 78.

183 *Paris Is Burning* (1990/1).

184 Ibid.; Hilderbrand, *Paris Is Burning*, p. 75.

185 *Paris Is Burning* (1990/1); Hilderbrand, *Paris Is Burning*, p. 82.

186 *Paris Is Burning* (1990/1).

187 J. Boles and K. W. Elifson, 'The Social Organization of Transvestite Prostitution and AIDS', *Social Science & Medicine*, 39:1 (1994), 85–93. However, the questions asked in this study were rather limited.

188 D. Valentine, *Imagining Transgender: An Ethnography of a Category* (Durham, NC, 2007), p. 114.

189 Ibid., p. 3.

190 Ibid., pp. 114 15.

191 Ibid., p. 117.

192 Ibid., pp. 117–19.

193 Ibid., p. 128.

194 Ibid., p. 131.

195 The quote is from *Salt Mines* (1990).

196 J. Cowen, 'The Dragqueen Stroll', unpublished typescript, 1989, New-York Historical Society Museum and Library.

197 S. Aikin, *Digging Up the Salt Mines: A Film Memoir* (New York, 2013), p. 8.

198 *Salt Mines* (1990).

199 Ibid.; Aikin, *Digging Up the Salt Mines*, p. 36.

200 *Salt Mines* (1990).

201 Ibid.

202 Aikin, *Digging Up the Salt Mines*, p. 10.

203 *Salt Mines* (1990).

204 E. P. Johnson, *Sweet Tea: Black Gay Men of the South* (Chapel Hill, NC, 2012), p. 349.

205 Ibid., pp. 339, 356–7.

206 Ibid., p. 405.

207 Ibid., p. 421.

208 Hwahng and Nuttbrock, 'Sex Workers, Fem Queens, and Cross Dressers', 43.

209 L. E. Kuper, L. Wright, and B. Mustanski, 'Stud Identity Among Female-Born Youth of Color: Joint Conceptualizations of Gender Variance and Same-Sex Sexuality', *Journal of Homosexuality*, 61:5 (2014), 714–31, quotes at 717, 724.

210 Ibid., 721.

211 Ibid., 724.

212 V. A. Rosario, 'Studs, Stems, and Fishy Boys: Adolescent Latino Gender Variance and the Slippery Diagnosis of Transsexuality', in C. Zabus and D. Coad (eds.), *Transgender Experience: Place, Ethnicity, and Visibility* (New York, 2014), ch. 4, quote at p. 57.

213 *The Aggressives* (2005).

214 Ibid.

215 Ibid.

216 Ibid.

217 V. Bailey, 'Brown Bois', *TSQ*, 1:1–2 (2014), 45–7; www.brown boiproject.org/blog.

218 J. Manion, 'Transbutch', *TSQ*, 1:1–2 (2014), 230–2, quote at 230.

219 J. Money and M. Lamacz, 'Gynemimesis and Gynemimetophilia: Individual and Cross-Cultural Manifestations of a Gender-Coping Strategy Hitherto Unnamed', *Comprehensive Psychiatry*, 25:4 (1984), 392–403.

220 J. Escoffier, 'Imagining the She/Male: Pornography and the Transsexualization of the Heterosexual Male', *Studies in Gender and Sexuality*, 12:4 (2011), 268–81.

221 D. Mauk, 'Stigmatized Desires: An Ethnography of Men in New York City Who Have Sex with Non-Operative Transgender Women', Columbia University Ph.D., 2008, p. 210.

222 Fleisher, *Drag Queens of New York*, p. 64.

223 M. S. Weinberg and C. J. Williams, 'Men Sexually Interested in Transwomen (MSTW): Gendered Embodiment and the Construction of Sexual Desire', *The Journal of Sex Research*, 47:4 (2010), 374–83, quotes at 378.

224 D. Valentine, 'Identity', *TSQ*, 1:1–2 (2014), 103–6, quote at 105.

225 Mauk, 'Stigmatized Desires', p. 118.

226 J. Ames, *The Extra Man: A Novel* (New York, 2010), p. 211. First published in 1998.

227 D. Mauk, A. Perry, and M. Muñoz-Laboy, 'Exploring the Desires and Sexual Culture of Men Who Have Sex with Male-to-Female Transgender Women', *Archives of Sexual Behavior*, 42:5 (2013), 793–803, quote at 797.

228 Weinberg and Williams, 'Men Sexually Interested in Transwomen', 382. See, also, M. S. Weinberg and C. J. Williams, 'Sexual Field, Erotic Habitus, and Embodiment at a Transgender Bar', in A. I. Green (ed.), *Sexual Fields: Toward a Sociology of Collective Sexual Life* (Chicago, 2014), ch. 2.

229 D. Operario and others, 'Men Who have Sex with Transgender Women: Challenges to Category-based HIV Prevention', *AIDS and Behavior*, 12:1 (2008), 18–26, quotes at 22.

230 Mauk, 'Stigmatized Desires', p. 98.

231 A. B. Tompkins, '"There's No Chasing Involved": Cis/Trans Relationships, "Tranny Chasers", and the Future of a Sex-Positive Trans Politics', *Journal of Homosexuality*, 61:5 (2014), 766–80, quotes at 772.

232 Ibid.

233 For example, A. Tompkins, 'Asterisk', *TSQ*, 1:1–2 (2014), 26–7.

234 Valentine, 'Identity', 105.

235 For example, W. A. W. Walters and M. W. Ross (eds.), *Transsexualism and Sex Reassignment* (New York, 1986), plates between pp.

102 and 103; D. A. Gilbert and others, 'Transsexual Surgery in the Genetic Female', *Clinics in Plastic Surgery*, 15:3 (1988), 471–87.

236 C. Griggs, *Journal of a Sex Change: Passage Through Trinidad* (New York, 2004), Foreword by J. Halberstam, p. vii. First published in 1996.

237 Ibid., pp. 211–12.

238 Ibid., p. 53.

239 S. J. Kessler, *Lessons from the Intersexed* (New Brunswick, NJ, 2002), pp. 66–8 (quote at p. 68). First published in 1998.

240 Ibid., p. 66.

241 A. Bolin, *In Search of Eve: Transsexual Rites of Passage* (New York, 1988), p. 177.

242 G. E. Teague, *A Social and Psychological Account of Gender Transition: The Diary of a Transsexual Academic* (Lewiston, NY, 2008), pp. 215–16.

243 A. Ashley and D. Thompson, *The First Lady* (London, 2006), pp. 123, 124.

244 For example, J. Jacques, 'Transgender Journey: Time for Sex Reassignment at Last', *Guardian*, 30 August 2012: www.guard ian.co.uk/lifeandstyle/2012/aug/30/sex-reassignment-surgery-transgender-journey.

245 Griggs, *Journal of a Sex Change*, p. 213.

246 T. T. Cotton (ed.), *Hung Jury: Testimonies of Genital Surgery by Transsexual Men* (Oakland, Calif., 2012).

247 N. J. McDaniel, 'A Phoenix's Quest for the Dragon', in ibid., pp. 102–9, quote at p. 106.

248 H. L. Talley, 'Facial Feminization and the Theory of Facial Sex Difference: The Medical Transformation of Elective Intervention to Necessary Repair', in J. A. Fisher (ed.), *Gender and the Science of Difference: Cultural Politics of Contemporary Science and Medicine* (New Brunswick, NJ, 2011), ch. 10.

249 E. Plemons, *The Look of a Woman: Facial Feminization Surgery and the Aims of Trans-Medicine* (Durham, NC, 2017), pp. 39–42.

250 For the former, see T. Rose, *My Facial Feminization Surgery* (Amazon Digital, 2005). For the latter, see Plemons, *Look of a Woman*, ch. 6.

251 Plemons, *Look of a Woman*, ch. 5: 'The Operating Room', quotes at pp. 121, 131.

252 McDaniel, 'A Phoenix's Quest for the Dragon', p. 107.

253 Plemons, *Look of a Woman*, p. 140.

254 CN Lester, *Trans Like Me: A Journey for All of Us* (London, 2017), Ebook, loc. 725.

255 Plemons, *Look of a Woman*, p. 85.

256 Ibid., p. 14.

257 C. Jenner and B. Bissinger, *The Secrets of My Life* (London, 2017), p. 293.

258 M. Haskell, *My Brother My Sister: Story of a Transformation* (New York, 2013), p. 112.

259 J. Mock, *Redefining Realness* (New York, 2014), pp. 135, 206.

260 Lester, *Trans Like Me*, locs. 730–5.

261 T. Newman, *I Rise: The Transformation of Toni Newman* (Santa Cruz, 2011), Ebook, loc. 934.

262 L. R. Murray, 'The High Price of Looking Like a Woman', *New York Times*, 19 August 2011.

263 K. Roen, '"Either/Or" and "Both/Neither": Discursive Tensions in Transgender Politics', *Signs*, 27:2 (2002), 501–22, quote at 505.

264 Ibid., 521.

265 S. Hines, 'What's the Difference? Bringing Particularity to Queer Studies of Transgender', *Journal of Gender Studies*, 15:1 (2006), 49–66, quotes at 63.

266 E. C. Davis, 'Situating "Fluidity": (Trans) Gender Identification and the Regulation of Gender Diversity', *GLQ*, 15:1 (2008), 97–130, quote at 103.

267 Z. Davy, 'Genderqueer(ing)', *Sexualities*, 22:1–2 (2019), 80–96.

268 J. Jacques, *Trans: A Memoir* (London, 2015).

269 J. R. Latham, 'Axiomatic: Constituting "Transexuality" and Trans Sexualities in Medicine', *Sexualities*, 22:1–2 (2019), 13–30.

270 Ibid., 26.

271 K. Amin, 'Temporality', *TSQ*, 1:1–2 (2014), 219–22, quote at 220.

272 J. R. Latham, '(Re)making Sex: A Praxiography of the Gender Clinic', *Feminist Theory*, 18:2 (2017), 177–204.

273 Ibid., 187.

274 G. Davis, J. M. Dewey, and E. L. Murphy, 'Giving Sex:

Deconstructing Intersex and Trans Medicalization Practices', *Gender & Society*, 30:3 (2016), 490–514. The research was carried out from 2008 to 2011.

275 Latham, '(Re)making Sex', 186.

276 Ibid., 188.

277 Ibid., 198.

278 N. Krieger, *Nina Here Nor There: My Journey Beyond Gender* (Boston, Mass., 2011), p. 188.

279 Ibid., p. 190.

280 R. Styles, *The New Girl: A Trans Girl Tells It Like It Is* (London, 2017), Ebook, loc. 3380.

281 Ibid., loc. 3385.

282 Ibid., locs. 3390 and 3416.

283 Ibid., loc. 3514.

284 M. Lovelock, 'Call Me Caitlyn: Making and Making Over the "Authentic" Transgender Body in Anglo-American Popular Culture', *Journal of Gender Studies*, 26:6 (2017), 675–87. See, also, B. Barker-Plummer, 'Fixing Gwen', *Feminist Media Studies*, 13:4 (2013), 710–24.

285 Jenner and Bissinger, *Secrets of My Life*, p. 102.

286 Lovelock, 'Call Me Caitlyn', 678.

287 Quoted in ibid., 683.

288 J. R. Latham, 'Trans Men's Sexual Narrative-Practices: Introducing STS to Trans and Sexuality Studies', *Sexualities*, 19:3 (2016), 347–68.

289 J. Rose, 'Who Do You Think You Are?', *London Review of Books*, 5 May 2016.

290 Latham, 'Trans Men's Sexual Narrative-Practices', 362. Latham is referring to trans men, but the description is applicable to trans people generally.

291 O. Gozlan, 'Introduction', in O. Gozlan (ed.), *Current Critical Debates in the Field of Transsexual Studies* (New York, 2018), pp. 1–12, quote at p. 3.

292 Bailey, *Butch Queens Up in Pumps*, p. 31.

293 P. B Preciado, 'Letter from a Trans Man to the Old Sexual Regime', *Texte Zur Kunst*, 22 January 2018: www.textezurkunst.de/articles/letter-trans-man-old-sexual-regime-paul-b-preciado.

294 K. E. Bryant, 'The Politics of Pathology and the Making of

Gender Identity Disorder', University of California Santa Barbara Ph.D., 2007.

295 J. Gill-Peterson, *Histories of the Transgender Child* (Minneapolis, Minn., 2018), quote at p. 5.

296 For a good introduction to the issues and disagreements in the treatment of children (and adolescents), see J. Drescher and W. Byne (eds.), *Treating Transgender Children and Adolescents: An Interdisciplinary Discussion* (New York, 2013), originally published in *Journal of Homosexuality* (2012).

297 L. E. Newman, 'Transsexualism in Adolescence: Problems in Evaluation and Treatment', *Archives of General Psychiatry*, 23:2 (1970), 112–21, quote at 117.

298 A. Travers, *The Trans Generation: How Trans Kids (and Their Parents) are Creating a Gender Revolution* (New York, 2018).

299 See J. Singal, 'How the Fight Over Transgender Kids Got a Leading Sex Researcher Fired', *New York Magazine*, 7 February 2016: www.thecut.com/2016/02/fight-over-trans-kids-got-a-res earcher-fired.html.

300 K. J. Zucker and S. J. Bradley, *Gender Identity Disorder and Psychosexual Problems in Children and Adolescents* (New York, 1995), esp. ch. 9.

301 K. J. Zucker, 'Children with Gender Identity Disorder: Is There a Best Practice?', *Neuropsychiatrie de L'enfance et de L'adolescence*, 56:6 (2008), 358–64, quote at 363.

302 G. Owen, 'Is the Trans Child a Queer Child? Constructing Normativity in *Raising Ryland* and *I Am Jazz: A Family in Transition*', *Queer Studies in Media & Popular Culture*, 1:1 (2016), 95–109, quote at 99.

303 C. Keo-Meier and D. Ehrensaft, 'Introduction to the Gender Affirmative Model', in C. Keo-Meier and D. Ehrensaft (eds.), *The Gender Affirmative Model: An Interdisciplinary Approach to Supporting Transgender Expansive Children* (Washington, DC, 2018), pp. 3–36, quote at p. 14. See, also, D. Ehrensaft, 'Realities and Myths: The Gender Affirmative Model of Care for Children and Youth', in Gozlan (ed.), *Current Critical Debates*, ch. 7.

304 T. Meadow, *Trans Kids: Being Gendered in the Twenty-First Century* (Oakland, Calif., 2018), p. 43.

305 P. T. Cohen-Kettenis, H. A. Delemarre-van de Wall, and L. J. G. Gooren, 'The Treatment of Adolescent Transsexuals: Changing Insights', *Journal of Sexual Medicine*, 5:8 (2008), 1892–7.

306 Hence the title of C. Castañeda, 'Developing Gender: The Medical Treatment of Transgender Young People', *Social Science & Medicine*, 143 (2015), 262–70.

307 J. Pyne, '"Parenting Is Not a Job . . . It's a Relationship": Recognition and Relational Knowledge Among Parents of Gender Non-Conforming Children', *Journal of Progressive Human Services*, 27:1 (2016), 21–48.

308 Travers, *Trans Generation*, p. 24.

309 S. Kuklin (ed.) *Beyond Magenta: Transgender Teens Speak Out* (Somerville, Mass., 2014), p. 106.

310 Ibid., p. 25.

311 Meadow, *Trans Kids*, p. 9.

312 M. Robertson, *Growing Up Queer: Kids and the Remaking of LGBTQ Identity* (New York, 2019), pp. 139–52.

313 See L. A. Saffin, 'Identities Under Siege: Violence Against Transpersons of Color', in E. A. Stanley and N. Smith (eds.), *Captive Genders: Trans Embodiment and the Prison Industrial Complex* (Oakland, Calif., 2011), pp. 141–62.

314 J. Tang, 'Contemporary Art and Critical Transgender Infrastructures', in R. Gossett, E. A. Stanley, and J. Burton (eds.), *Trap Door: Trans Cultural Production and the Politics of Visibility* (Cambridge, Mass., 2017), pp. 363–92, quote at p. 365.

315 R. Brubaker, *Trans: Gender and Race in an Age of Identities* (Princeton, NJ, 2016), p. 147.

316 P. Currah, R. M. Juang, and S. P. Minter (eds.), *Transgender Rights* (Minneapolis, Minn., 2006); P. S. Gehi and G. Arkles, 'Unraveling Injustice: Race and Class Impact of Medicaid Exclusions of Transition-Related Health Care for Transgender People', *Sexuality Research & Social Policy*, 4:4 (2007), 7–35; D. Spade, *Normal Life: Administrative Violence, Critical Trans Politics, and the Limits of Law* (Durham, NC, 2015); J. K. Taylor and D. P. Haider-Markel (eds.), *Transgender Rights and Politics: Groups, Issue Framing, and Policy Adoption* (Ann Arbor, Mich., 2015).

317 See D. W. Riggs, 'What Makes a Man? Thomas Beatie, Embodiment,

and "Mundane Transphobia"', *Feminism & Psychology*, 24:2 (2014), 151–71, quote at 167.

318 C. A. Shelley, *Transpeople: Repudiation, Trauma, Healing* (Toronto, 2008), p. 31.

319 M. Februari, *The Making of a Man: Notes on Transsexuality*, trans A. Brown (London, 2015), pp. 68–9.

320 For a useful summary, see M. K. Stohr, 'The Hundred Years' War: The Etiology and Status of Assaults on Transgender Women in Men's Prisons', *Women & Criminal Justice*, 25:1–2 (2015), 120–9.

321 C. Goring / C. R. Sweet, 'Being an Incarcerated Transperson: Shouldn't People Care?', in Stanley and Smith (eds.), *Captive Genders*, pp. 185–7, quote at p. 185.

322 V. Jenness and S. Fenstermaker, 'Forty Years after Brownmiller: Prisons for Men, Transgender Inmates, and the Rape of the Feminine', *Gender & Society*, 30:1 (2016), 14–29.

323 See L. Heidenreich, 'Transgender Women, Sexual Violence, and the Rule of Law: An Argument in Favor of Restorative and Transformative Justice', in J. M. Lawston and A. E. Lucas (eds.), *Razor Wire Women* (Albany, NY, 2011), ch. 12.

324 J. Sudbury, 'From Women Prisoners to People in Women's Prisons: Challenging the Gender Binary in Antiprison Work', in Lawston and Lucas (eds.), *Razor Wire Women*, ch. 14.

325 J. Sumner and L. Sexton, 'Same Difference: The "Dilemma of Difference" and the Incarceration of Transgender Prisoners', *Law & Social Inquiry*, 41:1 (2016), 616–42.

326 J. C. Oparah, 'Feminism and the (Trans)gender Entrapment of Gender Nonconforming Prisoners', *UCLA Women's Law Journal*, 18 (2012), 239–71, at 242.

327 L. Cannes, 'Transgender Murders: 2017 Off to a Horrific Start', *Lexie Cannes State of Trans*, 19 February 2017: https://lexiecannes com/2017/02/19/transgender-murders-2017-off-to-a-horrific-start.

328 https://lexiecannes.com/?s=murders&x=3&y=11.

329 TvT TMM Update: Trans Day of Remembrance 2018: https://transrespect.org/en/trans-murder-monitoring/tmm-resources.

330 https://tdor.translivesmatter.info/reports.

331 www.facebook.com/groups/TransViolence.

332 S. Lamble, 'Retelling Racialized Violence, Remaking White Innocence: The Politics of Interlocking Oppressions in Transgender Day of Remembrance', in Stryker and Aizura (eds.), *Transgender Studies Reader 2*, ch. 2.

333 Wright, *Trans/Portraits*, p. 83.

Conclusion

1 W. Bockting and others, 'Sociopolitical Change and Transgender People's Perceptions of Vulnerability and Resilience', *Sexuality Research and Social Policy*, https://doi-org /10.1007/s13178-019-00381-5. Published online: 19 February 2019.

2 B. Reay, *Sex in the Archives: Writing American Sexual Histories* (Manchester, 2019), ch. 5: 'A Transgender Story: The Diaries of Louis Graydon Sullivan'.

3 San Francisco History Center, San Francisco Public Library, Louis Graydon Sullivan Papers (hereafter Sullivan Papers), Box 2, Folder 52: Correspondence with C. D., 1987. D. was murdered in the 1990s: http://articles.courant.com/1993-02-15/news/0000105921_1_cabbie-union-station-beauty-school.

4 Sullivan Papers, Box 2, Folder 110: Transcripts of letters to and from 'Liz' [E. M.], 1975–8; Folder 109: Letters from Elliot/Dale [E. M.], 1980–1.

5 Ibid., Folder 110: Letter to E. M., 9 March 1976; Folder 59: Letters from G. G., 26 April, 2 June, and 23 July 1981; Folder 67: Letter to D. L., 18 September 1984.

6 Ibid., Folder 110: Letter to E. M., 9 March 1976.

7 Ibid., Folder 127: Replies to personal advertisement in the magazine *Coming Up!* 1984–5.

8 Sullivan Papers, Box 1, Folders 4–20: Sheila/Louis Sullivan Diaries and Journals, 1961–91 (hereafter Sullivan Diaries), 14 January 1980.

9 Ibid., 21 January 1979.

10 Ibid., 1 February 1980.

11 Sullivan Papers, Box 2, Folder 115: Letter to Virginia Prince, 7 February 1981.

12 Sullivan Diaries, 7 September 1981. Emphasis in original.

13 Ibid., 22 August 1981. Emphasis in original.

14 Ibid., 15 January 1984.
15 Ibid., 4 March 1984. Emphasis in original.
16 R. Brubaker, *Trans: Gender and Race in an Age of Identities* (Princeton, NJ, 2016), pp. 72–3. For the nonbinary, see M. Rajunov and S. Duane (eds.), *Nonbinary: Memoirs of Gender and Identity* (New York, 2019).
17 R. Spoon, 'Stories I Tell Myself and Others (Gender as Narrative)', in R. Spoon and I. E. Coyote, *Gender Failure* (Vancouver, 2014), pp. 239–44.
18 Ibid., p. 244.
19 J. Jacques, *Trans: A Memoir* (London, 2015), Ebook, loc. 655.
20 Ibid., loc. 1506.
21 Ibid., loc. 1099.
22 CN Lester, *Trans Like Me: A Journey for All of Us* (London, 2017), Ebook, locs. 1911–16.
23 Ibid., loc. 1921.
24 Ibid.
25 I am referring to R. C. Savin-Williams, *Mostly Straight: Sexual Fluidity Among Men* (Cambridge, Mass., 2017).

Index